ENVIRONMENTAL COMMUNICATION

Skills and Principles for
Natural Resource Managers, Scientists, and Engineers

Richard R. Jurin • K. Jeffrey Danter • Donald E. Roush, Jr.

Pearson
Custom
Publishing

Cover Photo: "Rock Wall in Forest," by Ken Shearer/Artville Stock Images.

Printed in the United States of America

10 9 8 7 6 5 4 3 2 1

Please visit our web site at www.pearsoncustom.com

ISBN 0–536–60833–4

BA 992086

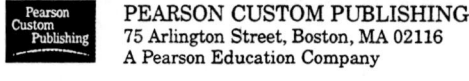

PEARSON CUSTOM PUBLISHING
75 Arlington Street, Boston, MA 02116
A Pearson Education Company

COPYRIGHT ACKNOWLEDGMENTS

CONTENTS

SUMMARY OUTLINE

Long Outline

PREFACE

On our cover you can see an old rock wall standing in a New England forest, built when the United States was a young nation. The rocks that were stacked together to make this barrier have become lichen-covered over the centuries.

Around the wall, a second-growth forest has replaced the virgin wood that was probably removed well before this fence was put in place. This structure likely marked the edge of a farm field. Its practical purpose was both to mark a boundary and provide a place for disposal of rocks, which made plowing difficult. Such a wall is the residue of difficult work.

The rock wall's communicative purpose was to proclaim dominion. The pioneers who built it meant to let others know that this was their land. A family probably made their living here. Theirs was a close relationship with the land.

This rock wall is, therefore, a message. It symbolizes the human-land interaction in the United States. It is an American environmental communication.

Photographer Ken Shearer captured this wall one autumn while touring the backwoods of Vermont and Maine. He recalls miles and miles of these rock walls standing along country lanes. Shearer uses large format cameras, which produce photographs of tremendous beauty and power. We liked this photograph very much and hope you do too.

This book is about communication and our environment. The cover of this book is meant to be indicative of what is inside.

We hoped to convey additional messages through the cover's graphic design. First, the earthy tones in the photograph are meant to remind viewers of the natural world. The greens, yellows, browns and grays within the image denote autumn in an eastern deciduous forest, an ecosystem familiar—at least vaguely—to most Americans. One shade of gray is picked up and repeated as the background color of the rest of the cover. Here our meaning was subtler. Environmental communication deals in many, many gray areas. Uncertainty is rife in environmental science; we do not have clear answers about the environmental effects of post-industrial human society. Message-makers will have to consider what is known and present remedies based on the preponderance of the evidence. This is not an easy task.

But, it is a necessary one if we as a society are to move toward sustainability. We are not alone in this type of thinking about environmental communication. As we were preparing the final manuscript for this book, we came across a letter in the journal Science. A distinguished group of ecologists, including many of whom we had read in depth during our graduate studies, wrote that the best science can no longer consist of just doing research and publishing findings in the technical literature. Now, scientists serving the public interest must add a third activity to their repertoire: "informing the general public (and, especially, taxpayers) of the relevance and importance of our work."[1]

Indeed, we believe this communication imperative applies to all professionals involved in the management of the Earth's natural heritage. Scientists, engineers, and natural resource managers must bring their knowledge and ideas into a larger public dialogue. As importantly, the populace will not necessarily accept as fact, or even respect the opinions of trained environmental professionals based just on reputation. In the modern world, communicated messages exist in a competitive marketplace of ideas and plain noise. Serious dialogue about the environment must cut through the clamor of this marketplace. In writing this book, we hope to make the point that environmental communication must be a planned, targeted process if it is to be noticed at all.

So, this is a book for anyone who communicates, whether for a non-profit organization, as part of a for-profit business, a government agency charged with managing a resource, a community needing to address an issue, or an individual. It is for anyone wanting to know what to do in a situation that demands more communications expertise.

As you go through this book, realize that even though we discuss communications in discrete "chunks," we hope you can see threads of integration among the chapters. It is our belief that one must always retain a holistic perspective on environmental issues. As John Muir said, "When we try to pick out anything by itself, we find it hitched to everything else in the universe."

The holism we have tried to incorporate in this text is a strategy toward the goal of sustainability—human habitation of our planet without compromising environmental quality, without destroying the other components of the biosphere.

If the goal is sustainability, then everything must be geared towards that goal. We hope that this book will give you knowledge and tools helpful for your own journeying as an environmental communicator. Whether you are an environmentalist, an industrialist, part of a community group, or an individual trying to make a difference, we hope this book offers a way to meet your objectives.

Good Luck.

No book is ever written in a vacuum and many people have influenced and helped the development of Environmental Communication. Each chapter was written by the authors, yet each is based on ideas generated by colleagues we know and respect. This book is truly a collaborative product.

ORGANIZATION OF THE BOOK

Throughout this book we have endeavored to keep the information to what we consider the essentials. This also keeps the book to a readable and manageable size. To this end, we have included a section of "Further Reading," at the end of each chapter, for those of you who desire to find out more.

The book is organized into three main sections. Section 1 sets the stage for why we need environmental communication. Chapters 2-4 emphasize essential aspects towards which any environmental communicator should be aspiring. This also includes the myriad audiences that the communicator may have targeted.

Section 2 covers the basics of planning. This is like orienteering yourself through a map of communications. If you know where you wish to go and how you intend to get there, you are more likely to arrive where you expected. Randomly setting out on a journey without a clear direction can be exciting, but, it can take you to destinations you would be better to avoid. Chapters 5-8 cover the essentials of planning your route to a successful communications strategy. Chapter 9 gives some advice about using some of the more commonly used mass media.

It is in Section 3 that we make you aware of the many communication skills you need to understand. Chapters 10–15 cover skills and applications you need when working directly with people. Successfully interacting with people requires a whole box of tools. These tools range from speaking dynamically to an audience, to understanding why they think and act the way that they do. The most prolific source of information today is the news media. Insights and help on dealing with this pervasive system will help communicators ensure that the information they give to the media is reported as it was intended to be heard. Chapter 17 gives insights and advice on how to manage conflict. This is a given in life, especially with environmental communications. Chapters 18 and 19 present tools in the communicator's repertoire that represent different strategies for a communications plan.

NOTE

1. This was part of a multiple-author letter to Science. Fakhri Bazzaz, Geraldo Ceballos, Margaret Davis, Rodolfo Dirzo, Paul R. Erhlich, Thomas Eisner, Simon Levin, John H. Lawton, Jane Lubchenco, Pamela A. Matson, Harold A. Mooney, Peter H. Raven. Joan E. Roughgarden, Jose Sarukhan, G. David Tilman, Peter Vitousek, Brain Walker, Diana H. Wall, Edward O. Wilson, George M. Woodwell. Ecological Science and the Human Predicament. Letter to the Science's Compass, Science, Vol, 282 (October 30), 1998.

ACKNOWLEDGMENTS

The impetus for writing this book was twofold. First, at a conference of environmental communicators in Chattanooga in 1995, a discussion arose about the need for a comprehensive and up-to-date introductory communications text for university undergraduate students and professionals in the field. Second, at The Ohio State University, a communications course is required for all students seeking a bachelor's degree from the School of Natural Resources. The packet of materials for that course evolved over many years, and this publication transfers this myriad of handouts, readings, ideas, and exercises into a unified text.

Although each chapter was written by the authors, it was in some way based on not just our ideas but on many other colleagues' ideas developed over several years. Their comments and feedback were truly invaluable. We specifically thank Dr. John Disinger, Dr. Rosanne W. Fortner, Dr. Joseph E. Heimlich, Dr. Gary W. Mullins, Dr. Robert E. Roth, and Dr. Stefan Sommer for ideas and comments with preliminary versions of the chapters. Dr. Nicholas J. Smith-Sebasto is gratefully acknowledged for his help with the preliminary draft of the outline and supportive comments.

We thank close friends and family members for their incredible patience and support. Richard especially thanks his wife, Sylvia, who put up with him during the long months of writing and made him play when he had worked too much and reminded him to write when he didn't want to.

Finally, to our editors over the long haul, Kurt Jaenicke and Dave Daniels, our grateful thanks for their support and patience over the several extended deadlines.

PREFACE

1

Understanding the World Around Us

We live on a unique and magnificent planet, a place of rare beauty and great value (Fortner 1991). And, every human on Earth crafts, exchanges, and receives messages about our home. We are all environmental communicators.

Each of us already partakes in the process that is the subject of this book—environmental communication. If we are already doing it, we can only hope to learn to do it better. More clearly. More effectively. In more meaningful ways.

As communicating is a skills-based process, we can learn to improve our abilities to send and decode information. Environmental communication, it can be argued, is based on an urgency to better understand and translate the human relationship with the rest of nature. As every person attempts to create meaning from the sensations produced by the world around them, our population grows with rapidity usually found only in the microbial realm. Human beings have a natural affinity to purposely work through their relationships with the rest of the natural world (Cantrill 1999). As we move toward the clarity we all seek, we like to discuss it with others. Our discourse about the world around us is environmental communication.

What Do We Know About Environmental Communication?

Our discussion of environmental communication rests on a foundation of ideas. Several axioms form this foundation.

Communication, as considered here, is a human activity. Clearly, scientists have identified many processes between non-human organisms that can be labeled "communication." But, the deliberateness and richness of messages that form environmental communication is only found among humans. Rightly, the assumption of humans somehow being above the rest of nature has been fingered as the source of many of our environmental problems. It is both ironic and hopeful that this distinction can now be used by environmental communicators to help overcome human-caused degradation of the biosphere.

You cannot not communicate.   Mere existence is an act of communication. To be is to communicate. Trying not to send out any messages sends a message in itself. If one sends messages regardless, it would be wise to communicate with purpose and competence. For those choosing the natural resource professions, this is doubly true. Being understood depends on properly forming and sending messages. Confusion is one likely result of poor communication.

Understanding is the goal of communication.   Communication is successful when a message is comprehended by its recipient. Many messages go uncomprehended, clouded by some glitch in the system. These glitches are called noise. Noise is to communication what entropy is to life, the thing against which the system constantly struggles. Noise can occur throughout any portion of a communication system. A communicator's goal is to overcome and circumvent noise.

Most responsibility in this process rests with the communicator, not the recipient of the message.   Carefully and skillfully, successful communicators package their messages for maximum effect. They should know to whom they are sending information, how this audience prefers to receive such information, and how they can be expected to translate it. Knowing why one wants to send messages helps a communicator shoulder this responsibility. So, ethics play a decisive role in environmental communication. When a message is not understood, the fault falls back on the originator of the message, the communicator.

Human society depends on nature for survival.   Everything we do within our highly developed and specialized human society depends on the services provided to us by a living, healthy planet (Baskin 1997; Daily 1997). Earth is the only home we have and the functions of its biosphere sustain us. Natural systems give us clean air, clean water, food, shelter, pleasure, beauty and belief in affairs beyond ourselves. Further, our economy—the human institution most important in political deliberations around the world—depends on nature's economy. In short, we live and work if, and only if, nature lives and works.

The Earth has its own messages to share with us.   Listening to the planet is one way of stating the work of science. Scientists are a crucial source of information for all environmental communicators. The position of environmental communicators between scientists and the larger non-technical population is precarious and yet exciting. The translation of scientific findings provided by environmental communicators is also vital. Those practicing environmental communication will need to hone their own perceptual skills, to understand what the planet has to say through them. Human senses have been extended by all sorts of gadgetry and instrumentation. We would understand much less about the state of Earth and the life it sustains without this technology. But, many commentators contend that our technology has blocked vital messages coming from the planet itself. Technology, the stuff of modern life, becomes noise in this interpretation. Though we are not the originators of these planetary messages, we may wish to learn to mute this particular distraction. Environmental communication can help. Environmental communicators can, too.

Atop the basis provided by these presumptions, a conceptual framework has been constructed. From this framework can be drawn a myriad of skills useful to environmental communicators. These principles and skills of environmental communication are examined in this book. To be most useful, a book should contain principles and skills that can be applied immediately. We hope you find practical suggestions in this text that you can apply in your work, for our world.

A BRIEF HISTORY OF ENVIRONMENTAL COMMUNICATION

Environmental communication arises from ideas as old as human interaction with each other and nature. But environmental communication also has a history that can be meaningfully traced to help understand the trails that led us to this place in our practice of it.

Contemporary American environmental communication is rooted at least as far back to the late 1800s. Pioneers in examining the American interaction with the land and waters of the country have names you've probably heard before: Henry David Thoreau, John Muir, Fredrick Law Olmstead, George Perkins Marsh, John Wesley Powell, Gifford Pinchot, Stephen Mather, and Aldo Leopold. This progression of men encouraged Americans to think in environmental ways. Their writings are bulwarks of the environmental history of the United States and have been influential worldwide.

Their perspectives did not always agree, yet they all share one similarity—their subject was the environment in its entirety. Throughout, a theme declared natural resources must be carefully conserved if we are to continue to thrive. In the early years, there was more of an emphasis on conservation of natural areas and prudent use of natural resources. Leopold is credited with establishing an ecological connection in our thinking and acting holistically within our environment.

Looking at these antecedents of today's environmental communication magnifies our appreciation of today's situation. Schoenfeld (1981) states, "Irrespective of their roots, [there are] common denominators among the various forms of environmental communication . . . All are focused on a comprehensive rather than a compartmentalized approach to the people-resources-technology system. A basic theme in environmental communication hence is interdependence—that everything is connected to everything else." More recently, Rachel Carson, Stewart Udall, Wallace Stegner, and Edward O. Wilson have carried on this tradition. Schoenfeld (1981) noted five roots that ground environmental communication, using environmental magazines as examples.

Nature Writing

One of the oldest types of writing, this genre dates back to ancient times when writers commented on newly-discovered lands and extreme weather occurrences. Early American writers like Muir and Thoreau wrote extensively of nature's acts. Nowadays nature writing can be found in the magazines of many organizations. Magazines like *Field and Stream, Audubon,* and *Outside* are just some of the many that communicate about nature. Many TV documentaries found on Public Broadcast Service deal with nature with the same underlying sense of awe and detachment found in the writings of ancients.

Outdoor Recreation and Travel Writing

This form of writing is closely aligned with nature writing. Indeed, many environmental magazines will cover nature, outdoor recreation, and travel writing in the same issue, whereas others remain specialized to specific types of outdoor sports or travels. Early travel writers are credited with spurring on the mass emigrations to the United States. The journals and writings of the early pioneers who trekked across the Great Plains and Rocky Mountains toward new lands riveted readers on the east coast and in other countries.

Science Writing

Many of the premiere science journals and magazines began publishing in the mid-1800s. *Scientific American* first appeared in 1845 and *Science* in 1883. The American Chemical Society News Service began in 1919 and *Science News* in 1922. In addition, the land grant university system, created in 1887, is credited with the proliferation of "technology transfer" to farmers via farm journals and extension service publications. The reporting of science is not without controversy, as science has grown more specialized and esoteric. Environmental communicators face nearly insurmountable obstacles in producing clear and concise explanations of scientific findings. Even in 1973, James Larsen lamented that modern media coverage of science "falls far short of what could be considered adequate, considering the critical importance of an informed electorate on matters of significance to modern civilization."

Public Affairs Reporting

Newspaper reports of government and business activities, the backbone of print journalism, dates to the "muckraking" of the late 1800s and early 1900s. This form of journalism was meant to protect the public good.

Reporters are advocates of the "little guy" up against the Goliaths of industry and other oppressions. They see themselves as giving a voice to the voiceless. New technologies and more urbanized populations allowed the news media to become a powerful force of reporting of life everywhere. While gatekeeper constraints are still an everyday reality of the news media, the first national newspapers and now the broadcast news media are still the primary mode in which most people find out about the world beyond their own senses and what is happening.

Persuasion

Persuasive writing can be traced back to the Middle Ages. Consider Martin Luther who pinned his challenge statements to a church door. Within the last century, Gifford Pinchot, as head of the new U.S. Forest Service, used persuasion to get Congress to set aside vast areas of the United States as national forests. The modern national parks, designated wildernesses, wildlife refuges, and national monuments are also related to his success, since most were carved out of the National Forest system. In more modern times, David Brower created several non-governmental organizations that collectively were able to persuade Congress to enact many laws to protect the environment. These groups and many of their offshoots still actively lobby Congress, publish magazines and other literature, and develop programs to help the environment. Other groups in opposition have also set up persuasive campaigns to counteract these forces. The power in persuasion seems to be with those that appeal to the consciousness of the population and the prevailing zeitgeist.

THE COMMUNICATING INFORMATION MODEL

Visualizing the entire system of environmental communication, while an imperfect exercise, can assist one in understanding the variety of influences and roles played by actors who disseminate and consume information about nature and natural resources. The Communicating Information Model[1] (see Figure 1), while not the most simplistic depiction available, provides a useful and accurate picture of this system. The arrows represent major information flow within the model. Emphasis of information flow is represented by the strength of the line. For instance, a heavy line signifies a major flow of information while a dashed line indicates a minor channel or somewhat restricted flow of information.

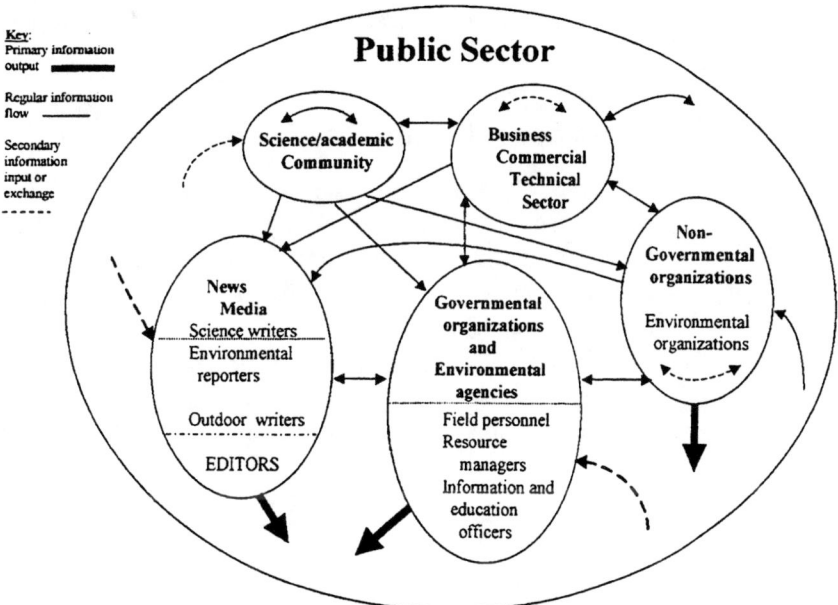

Figure 1   The Communicating Information Model

Information Sharing Within and Among Specific Groups

Consider how information sharing and communication occurs within and among various groups listed on the Communicating Information Model. Different groups will have different missions and goals, but generally within society these interactions will hold true.

Scientific Community

This sector tends to be the source of most reliable information in society. Many societal decisions are based on scientific information. But with today's increase in use of technology and occurrence of environmental problems, information has little time to be discussed within the sector before it is used by the other sectors.

Hence, reliability of information is now a prime problem. Open communication mainly occurs through use of scientific journals and open scientific conferences; consensus is gained over time. The primary audience is not the public sector but the rest of the science community. Scientific Community to Business/Commercial/Technical sector is often about collaborative efforts in research. Basic research is expensive to set up, and yet many research labs in academic institutions have facilities and expertise to do such research, yet lack adequate funding. It is easier for many businesses to fund such set-up labs in exchange for specific research information. The public affects the scientific community in a tertiary way, by dictating the major kind of research interests that can be studied. Various grant & funding sources are more susceptible to public opinion and demands, especially public organizations. The News Media and Governmental Organizations and Agency sectors will review new findings in the general media or be alerted to new results by the science community. Alternatively, these two sectors will seek out potential experts in order to guide decision-making processes. The non-governmental organization sector will tend to use information that has filtered through the mass media. Understanding scientific communication is covered in Chapter 3.

Business/Commercial/Technical

Competition creates a source of communication. Communication within the sector tends to be more intragroup (within individual organizations) and limited among different groups. Since the organizations tend to be oriented towards financial goals, there is less cooperation and information sharing among groups because ideas are potential money makers. Primary external audiences are the public in their role as consumers, and regulatory agencies. Purchasing is a form of feedback. The public also provides feedback in the form of complaints or praise (through mail and phone calls to individual businesses) about company products, philosophies, or procedures. This business/commercial/technical sector also interactively communicates with the governmental organizations to fulfill legal requirements.

News Media

The primary role of the news media is reporting and analysis of current events, often prompted by competitive spirit (i.e., not to be "scooped"). The primary audience is the public sector. Secondary interactions will be with the agencies and science community who are a source of information and potential stories. The public will have a smaller feedback role by alerting the main news media to potential stories and providing background for current events. Smaller news outlets and the social sections of many major news media depend more on public feedback to alert them to more local news. News media are addressed in Chapter 16.

Governmental Organizations and Environmental Agencies

Legislation (laws, rules and regulations) requires compliance and enforcement action between the agencies and the business/commercial/technical sectors who are the primary audiences. The public sector is also a primary audience for outreach and education efforts through mass media. The public sector also has a smaller feedback role by alerting agencies and legislative bodies to public concern.

Non-Governmental Organizations (NGO's)

NGO's develop a whole persuasive communication network of their own. Primary audiences are the public sector through mass media, and legislative bodies of the government primarily through lobbying. While many groups within this sector may interact regularly, some groups tend to be isolationist. Each may focus on persuading the public and legislature to act on a focused interest. Secondary audiences will be the business sector, which can provide funds and political support as well as being the focus of action to modify business behavior. The news media are a means of expanding situation, event and organizational awareness. The public will also have a secondary interaction level through letter writing, phone calls and financial support (i.e.: contributions, membership, and subscriptions). Fund raising is a primary emphasis of many NGO groups because of their non-profit nature. Examples of these kinds of groups are Mothers Against Drunk Drivers (MADD), The Sierra Club, Oxfam International, and The National Rifle Association (NRA).

REFERENCES AND FURTHER READING

Baskin, Y. (1997). *The work of nature: How the diversity of life sustains us.* Island Press.

Cantrill, J. (Summer 1999). The environmental communication commission. *Ecologue* (newsletter of the Environmental Communication Commission of the National Communication Association). p. 1.

Daily, G. C. (1997). *Nature's services: Societal dependence on natural ecosystems.* Island Press.

Fortner, R. W. (ed.) (1991). *Special earth systems education issue of science activities,* 28, p. 1.

Larson, J. (1973). Science, communications, society. *The Journal of Environmental Education,* 5(1): 21–22.

Schoenfeld, C. (1981). *The environmental communication ecosystem: A situation report.* SMEAC information reference center/ERIC Clearinghouse for Science, Mathematics, and Environmental Education, The Ohio State University, Columbus, OH, pp. 11–15.

NOTE

1. Communicating information model—expanded and adapted from an original idea by William Witt, (1973), *The Journal of Environmental Education,* 5(1), 58–62.

SECTION 1:
PRINCIPLES OF ENVIRONMENTAL COMMUNICATION

2

COMMUNICATING ABOUT THE ENVIRONMENT

When a natural resource professional says "environmental communication," what do they mean? The concept couples two terms, both of which are probably familiar. Still, both of them encompass large areas of meaning. Attempting to state concise yet conclusive definitions for "environmental" and "communication" runs a high risk of failing. But, not defining one's terms carries an even higher risk of causing confusion later.

First, by "environmental" we defer to a nearly poetic definition put forth by Schoenfeld (1969), with input from a panel of ecologists. Here are their rhythmic, colorful phrases:

> *In locus, the fouled, clogged arteries of the city quite as much as scarred countryside.*
>
> *In scope, a comprehensive, interrelated humankind-environment-technology system.*
>
> *In focus, global environmental impacts of crisis proportions threatening the well-being of all humankind on an over-crowded planet.*
>
> *In content, tough ecological choices, not easy unilateral fixes.*
>
> *In strategy, long-range impact analyses and rational planning.*
>
> *In tactics, grass-roots participation in resource policy formation—in the streets and through institutional channels.*
>
> *In prospect, a necessary reliance on alternative sources of energy.*
>
> *In philosophy, a commitment to less destructive technologies and less consumptive lifestyles.*
>
> *In essence, a recognition of pervasive interdependencies, that everything is connected to everything else.*

13

As for "communication," the most popular American dictionary, the Merriam-Webster, gives this meaning: "a process by which information is exchanged between individuals through a common system of symbols, signs, or behavior." Communication seems the more intuitive concept of these two terms. But definitions only take you so far in grasping a concept.

Understanding the process of communication can be enhanced through the use of models—graphic representations of some phenomenon. Numerous models explain how messages are sent and received and explain the many problems that can occur. The simplest model shows a sender selecting a channel through which a message is then transmitted to a receiver. For example, say you need to get a message to a fellow dormitory resident and decide to walk down the hallway and talk to her. You decided here to use face-to-face communication, with your voice and sound waves as your channel. Alternatively, you might have chosen to use a telephone or e-mail as your channel. In any case, the message you sent would be almost the same. Information would be transferred from you to your colleague.

This simple model addresses only one-way communication, however. It fails to take into account the dynamics of two-way communication, where senders and receivers switch positions within the model as they take part in dialogue. Such dialogue might be an interpersonal conversation, a telephone call, exchange of letters, trading e-mail messages, or any other of a number of possible interchanges. The inherent alternation of roles in two-way communication produces another important feature within the system—feedback.

Feedback involves a communicator's review and evaluation of message receipt and decoding. Feedback permits messages to be improved, whether by selection of a new channel or modifications within the message. Even if the communicative interaction is not in real time (as it is with mass media), the sender should strive to ascertain if the message was received and clearly understood.

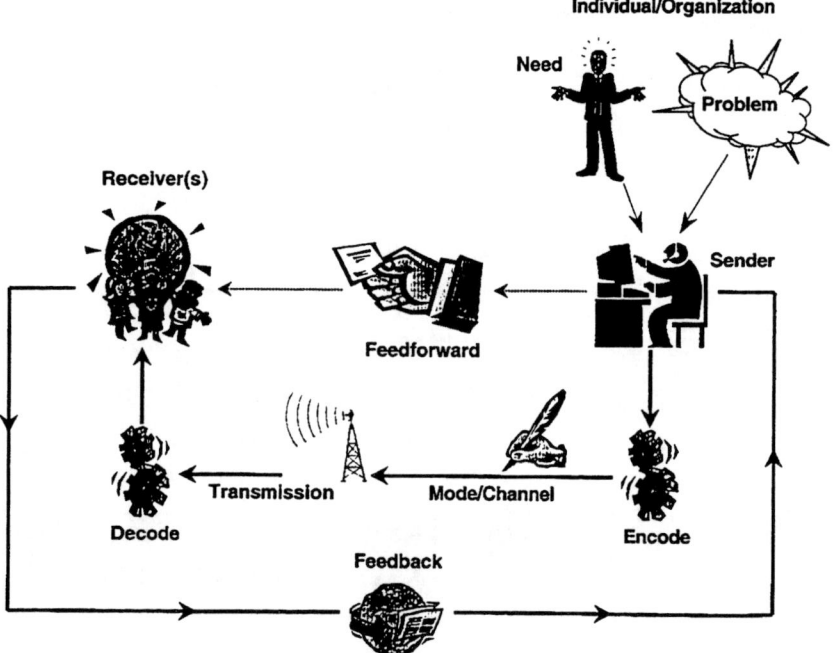

Figure 2a   The Communication Model

The cyclic model presented here depicts a dynamic system with sender and receivers interacting, each in turn encoding, transmitting and decoding. Thus, the sender becomes the receiver and the receiver becomes the sender. As can be seen in this model, a need exists for the sender to send a message, and to encode that message, in a form likely to be understood by the receiver. A mode/channel is selected and a message sent. This message is mentally processed by a receiver, who then responds in some manner. A receiver then alternates and becomes sender. The system, thus, perpetuates itself. Or, more correctly, communicating is an ongoing process.

Any communication process is imperfect, as noise permeates the entire system. This pervasive impurity works against comprehension of messages. Noise may be sounds unrelated to the message, but it can also be non-sounds. For instance, using engineering jargon in writing a children's book, providing a CD-ROM to a person without a computer, and showing a painting done only in subtle shades of pastel red, blue, and green to a color-blind person all would introduce lethal amounts of noise into a communication. The message would not get through. Anything present within a communication system that works against understanding is called noise. Communicators strive to reduce it.

Closely related to feedback and helpful in overcoming noise, feedforward allows communicators—both senders and receivers—to anticipate message exchange. You've probably experienced preparing for a complex session of communication. Studying for a test is one example. Deciding to see a movie based on a preview is another. When a sender forwards a simple prompt to ascertain if a receiver is willing and able to accept a message, this is feedforward, too. Feedforward allows a communicator to plan the expenditure of time and effort used in developing and interpreting messages. Mark Twain is credited with having once said, "Tell 'em what you're going to tell 'em; tell 'em; tell 'em what you told 'em." The first part of the adage is feedforward.

In examining models, there is a tendency to view the components as fixed and static, rather than as parts of a dynamic process. It is important to realize communication involves continually switching and changing of the sender-receiver roles. The model presented here is meant to guide you through the basic steps of the send-receive-repeat routine. Planning attractive and effective environmental communication relies on awareness of these modalities.

What's the Difference Among Environmental Communication, Environmental Education and Environmental Interpretation?

To this point, we've considered "communication" as a basic function of being human. And, we've developed a concept of "environmental communication" as evolving from an instinctual desire we have to understand the world around us. We create meaning from sensations and then converse about these meanings with others. Within this domain of endeavor, three related and overlapping fields have been established: environmental communication, environmental education, and environmental interpretation.

Try not to be confused by the redundancy of having environmental communication as the overarching concept, as well as one of the three fields within. "Communication" covers a lot of conceptual territory. We contend that you must communicate to educate and to interpret. But, even though it is tautological, you must communicate to communicate, too. Let us explain more.

The fields of environmental communication, education, and interpretation (ECEI) can be likened to a mythical monster, Cerberus the canine guardian of Hades, which had three heads atop three necks protruding from a common body. The heads, though able—at least in legends—to attack different targets, were essentially the same creature. A natural resource professional can anticipate needing to call on communicative, interpretive, and educational skills daily in their on-the-job tasks.

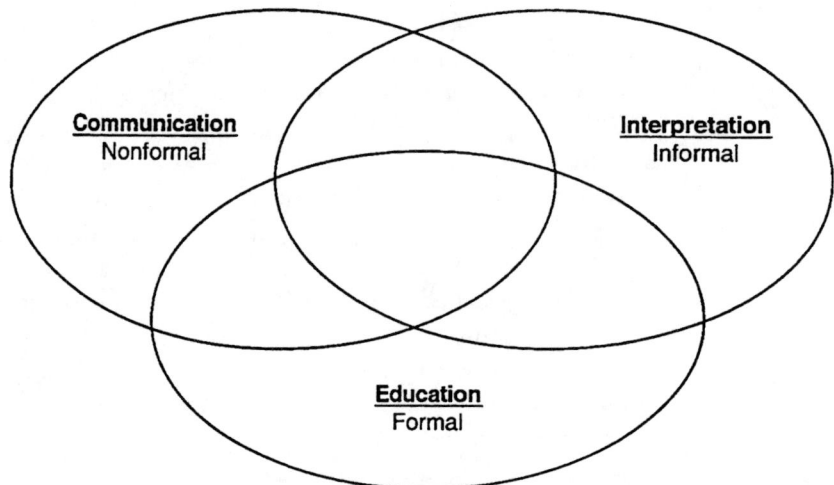

Figure 2b   Overlapping relationships of environmental communications, education, and interpretation, for the professional

So, the terms environmental communicator, interpreter, and educator can, for most intents and purposes, be used interchangeably. Yet, there are distinctions between the three. The following discussion delineates among the three fields, while also emphasizing their common features. A handy way to distinguish between environmental communication, education, and interpretation uses institutional setting of the informational exchange and audience focus to sort between formal, informal, and nonformal (Mocker & Spear, 1986).

Formal education, by and large, takes place inside classrooms. Classroom-based institutions, from pre-schools through universities, are formal educational centers. In formal settings, students are usually evaluated on criteria based in specific lesson objectives. Learning is achieved through interaction with a teacher. Although some relaxed teaching methods may permit substantial interactive dialogue among teachers and students, learning is highly structured in formal settings. Most often, learners submit to the educational goals of the institution and are required to attend formal education classes. They are a captive audience. The educational institution holds the decision-making power over what is learned and where learning takes place. *Environmental education occurs in formal settings.*

Informal education features less structure, no evaluation (no required tests of knowledge), and no requirement of attendance by learners. Learners attend because they are interested or desire entertainment. Institutions of informal education include museums, parks, nature centers, wildlife refuges, zoos, aquaria, art galleries, and historical sites. Learners can elect to attend informal educational programs when visiting an informal educational facility. Learning is often facilitated by a professional, perhaps a naturalist, docent, or living history actor. Interaction is loose compared to formal education, and learners are encouraged to ask questions freely. There exists for the visitor a personal option to interact with the interpreter. Informal learning needs to be entertaining because it is a leisure-time activity for visitors. The educational institution decides what is learned about, but cannot dictate whether learners attend to the lessons being offered. *Environmental interpretation occurs in informal settings.*

Nonformal education involves information disseminated primarily through mass media—television, radio, newspapers, pamphlets, fact sheets, billboards, magazines, the World Wide Web, etc. These communications can transcend place and time. Even though you've never actually been to the Amazon rainforest, the Siberian steppe, or Antarctica, you probably still have some idea what each looks like and the environmental issues that affect these places. Messages in the mass media provided you with that information. Learning via mass media is controlled by the learner, who decides unilaterally what to pay attention to. Nonformal learning is also a leisure-time activity, though it may carry undertones of requirements for job or school. By picking and choosing what mass media messages to subscribe to and commit to memory, the learner controls both what is learned, when to learn it, and where learning takes place. They are only limited by the amount of information given out by the medium to which they pay notice. *Environmental communication occurs in nonformal settings.*

Each of these three settings—formal, informal, and nonformal—depends on a mediator. This mediator may be called a teacher, facilitator, host, interpreter, journalist, communicator, tour guide, information/outreach specialist, educator, scientist, engineer, or one of many other titles. The environmental communicator's job is to transmit environmental, scientific, or natural resource information to interested recipients.

A fourth type of setting is possible. **Self-directed learning,** where the learner discovers information through first-hand experience, occurs outside of ECEI. This is because self-directed learning is not within the control of any mediator or institution. No professional helps the learner in any way in self-directed learning.

In Figure 2b, the overlapping circles signify that communicators/educators/interpreters will find themselves involved in doing two or even all three types of activity as the situation commands. Their audiences have subtly different motivations which will help a professional pick techniques from environmental education, interpretation, or communication. The techniques and resulting ideas can be ascribed to each, any, or all of the components. In many cases the distinction between the components may be difficult to see.

Implications for the Professional

Communicators, interpreters, and educators need to understand the context in which they hope to transmit their environmental information. A single topic will have to be developed and packaged in different ways for different audiences. To summarize, here's a final splitting of these fields:

Environmental Education

Mode of education	Formal—tightly structured learning with specific objectives
Main media	Personal presentation, structured interaction
Main focus	To teach
Primary interaction	One-on-group
Audience	Not volunteers
Main institution	Schools

Environmental Interpretation

Mode of education	Informal—loosely structured learning/augmented incidental learning
Main media	Personal narration, interactive dialogue, interpretive signage/displays (exhibits, trails, etc.)
Main focus	To engage the visitor
Primary interaction	One-on-one, or one-on-small group, or directioned self-guidance.
Audience	Volunteers
Main institutions	Historic centers, museums, wildlife refuges, art galleries, zoos, museums, parks, nature preserves, etc.

Environmental Communication

Mode of education	Nonformal—incidental structured learning
Main media	Mass media (newspapers, magazines, brochures, pamphlets, fact sheets, displays, TV, radio, Internet)—transcends place and time
Main focus	To inform
Primary interaction	Through audience size and reaction to advertising—purchasing power
Audience	Volunteers
Main institutions	Mass media companies

Regardless of all the parsing we could do to differentiate between environmental communication, education, and interpretation, they remain so closely related that if they were organisms they would certainly be classified as the same species. Consider some of the largest professional organizations of environmental communication professionals:

National Association for Interpreters (NAI)
International Association of Public Participation (IAP2)
Society of Environmental Journalists (SEJ)
North American Association for Environmental Education (NAAEE)

NAI has an environmental education section. IAP2 has members in academia, industry, government, as well as free-lance writers and artists. NAAEE has a section devoted to different groups of nonformal professionals. SEJ allows educators to become members. It's fair to say environmental professionals dealing with communication are not hung up on labels for themselves.

REFERENCES AND FURTHER READING

Fazio, J. R., & Gilbert, D. L. (1986). *Public relations and communications for natural resource managers*. Kendall/Hunt Publishing Co.

Ham, S. H. (1992). *Environmental interpretation*. North American Press.

Hartman, L. A., & Hanna, J. D. (1987). Interpretive educational courses in the United States and Canada. *The Journal of Environmental Education,* 18(4), 1–7.

Heimlich, J. E. (1993). *Non-formal environmental education: Towards a working definition.* ERIC Clearinghouse, Columbus, OH.

Jacobson, S. (1999). *Communication skills for conservation professionals.* Island Press.

Kleis, R. J. (1974) Non-formal education: The definitional problem. *Program of Studies in Non-formal Education Discussion Papers Number 2.*

Machlis, G. E., & Field, D. R. (Eds.) (1984). *On interpretation: Sociology for interpreters of natural and cultural history.* Oregon State University Press.

Mocker, D. W., & Spear, G. E. (1982). *Lifelong learning: formal, non-formal, informal, and self-directed.* ERIC Clearinghouse, Columbus, OH.

Monroe, M. C. (Ed.). (1999). *What works: A guide to environmental education and communication projects for practitioners and donors.* New Society Publishers.

Mullins, G. W. (1984). The changing role of the interpreter. *The Journal of Environmental Education,* 15(4), 1–3.

Parker, L. (1997). *Environmental communication: Messages, media & methods.* Kendall/Hunt publishing Co.

Radcliffe, D. J., & Colletta, N. J. (1989). Nonformal education. In *Lifelong education for adults: An international handbook*, Titmus, C. J. (Ed.), pp. 60–63. Pergamon Press.

Schoenfeld, C. A. (1969). What's new about environmental education? *The Journal of Environmental Education,* 1(1), 1–4.

3

DEVELOPING YOUR ENVIRONMENTAL LITERACY

Environmental literacy denotes an individual's set of abilities and commitments necessary to find, understand, assess, and act on information about the health of our environment. So, environmental literacy embodies values, beliefs and attitudes toward sustaining a healthy environment. Prerequisite to environmental literacy is a standard conception of literacy and a more specific idea, science literacy. An environmentally literate person understands the workings of modern science, and of policy-making. They also know how to apply their abilities to affect changes in society. Each of these elements builds on the others. Being aware of and having knowledge about environmental problems only supports the higher order skills necessary to be fully environmentally literate.

To understand environmental literacy, we must first understand literacy and, then, science literacy. Literacy encompasses some form of competency, whether in literature, cooking, yachting, child rearing, playing the violin, or any other recognized vocation. At the core of literacy is the ability to read and write so as to learn from the knowledge of others and then contribute to any particular body of human endeavor.

A person communicating environmental information needs to be able to judge accurate and relevant scientific information and relate it in a non-biased and credible way to a broader, non-scientific audience. Likewise, to be able to judge what human activities are sustainable requires the communicator to know about the functioning of the environment, to be sensitive to extra ecological pressures placed on the environment by humans, and to know what constitutes wise decisions. The communicator then attempts to help others go through the same thought processes. By doing that, environmental communicators assist others in becoming environmentally literate. This chapter discusses literacy, science literacy, environmental literacy, and, finally, the arts of filtering knowledge and argumentation. Knowing how all these notions interact informs the communication process. You'll then be better able to develop clear and credible messages for all the audiences you'll need to reach as an environmental communicator.

LITERACY

Being able to read and write is indispensable for citizens seeking to function well in modern societies. The new service-based economies of post-industrial developed nations place greater value on high levels of literacy. This implies a continuum along which an illiterate person gains skills in reading and writing, and subsequently may attain a threshold level that allows them to be literate enough to contribute to society. In a western society, this means people are comfortable with certain functions such as being able to read and fill out a tax document, read and understand the main parts of a newspaper, understand product labeling, and comprehend the directions on a drug prescription. Young children and adults who never learned to read and write would be at the lower end of the continuum. Toward the upper end would be people who are the most gifted readers, writers and critical thinkers.

Miller (1989) reported that one-quarter of the U.S. populace was either illiterate or functionally illiterate. If valid, this finding presents a huge problem for any communicator when developing any message for non-specialist audiences in the United States. The U.S. census defines illiteracy as anyone above the age of 14 that possesses less than a fifth-grade education. People with literacy deficiencies have difficulty coping with common day-to-day chores. Consider:

- Twenty percent cannot read a bus schedule or address a letter.
- Understanding a typical insurance policy requires a 12th grade reading level.
- Reading the instructions on over-the-counter medicines takes a 10th grade reading level.
- Making a TV dinner or filling out a tax form requires an 8th grade reading level.
- Comprehending the data on a driver's license uses a 6th grade education.

Remediation of reading problems is the focus of many in-school and adult education programs. About 10 percent of those 65 million Americans who lack literacy are enrolled in some form of remedial education program. That's a start, but it is not a final solution to allowing all to participate fully in the American dream of democratic participation in society.

SCIENCE LITERACY

Science is a complex subject, which requires a special set of analytical skills to understand. In that scientific findings play a large role in policy-making, scientific literacy is important for democratic functioning. To attain science literacy, citizens need to (Miller 1989):

- Understand the vocabulary of science.
- Understand the scientific method, the process that differentiates science from other ways of knowing.
- Comprehend research's iterative practice that views knowledge as tentative for centuries.

- Possess critical reading skills which enable one to judge valid and reliable scientific findings from incorrect, biased, or misleading ones. Part of this requirement is an understanding of the peer-review process, wherein research reports are reviewed by other scientists before being published or presented publicly.

- Analyze the costs and benefits of science and technology on the global village.

- Realize each new piece of knowledge adding to the overall picture of the universe's function. Therefore, few studies are ever really "breakthroughs."

- Further realize that uncertainty is always present in the scientific knowledge. So, it should always be reported.

Science is complex. Even experts rarely have a foolproof answer to any problem. Understanding of natural phenomena is imperfect. Science seeks to build consensus, not prove this or that. Science is like an ever-expanding jigsaw puzzle: as pieces are found and the picture nears completion the borders expand. Therefore, new information is constantly required (Bauer 1994).

The relationship between standard literacy and science literacy can be envisioned as the intersection of two continuums. In the diagram, the horizontal line represents literacy, one's reading and writing competency. The vertical line signifies science literacy.

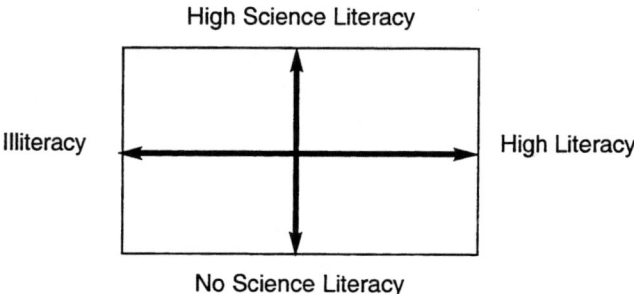

Now look at the four quadrants formed by the lines. Individuals and groups of people can fall into any of the four quadrants. An excellent communicator would be able to craft message which could be understood by persons in all of these categories. Be aware that the type of message should by necessity be different for each quadrant audience, and the medium may differ to reach each quadrant. (We'll analyze audiences in later chapters.)

ENVIRONMENTAL LITERACY

The building of modern environments consisting of cities, highways, agricultural fields, and factories has increased human potential to affect planet-wide ecology. We now have the power to degrade our entire biosphere. Environmental literacy acknowledges this power and saddles it with responsibility to focus on restoration, conservation, and sustainability. This responsibility involves all human discourses about our inter-relationships with the environment. An environmentally literate person is able to perceive and interpret the relative health of environmental systems and to take appropriate action to maintain, restore, and improve the health of those systems. A prerequisite to environmental literacy is being scientifically literate. Orr (1992) notes that environmental literacy carries within it a quality of mind that seeks connections and draws from intimate experience with at least one ecosystem. So, working knowledge of ecology is critical for anyone hoping to make good decisions about sustaining the environment.

The concept of environmental literacy implies (Roth 1992):

Knowledge of environmental issues and the credible science behind them.

Understanding of the "whole picture" and not just minor parts of it.

Empathy toward the total environment.

Knowledge of action skills.

Environmentally responsible beliefs, values, and attitudes.

Willingness to invest personally.

Active involvement in solving environmental issues.

Degrees of Environmental Literacy

As with literacy and science literacy, there are various magnitudes of environmental literacy. That is, environmental literacy also lies along a continuum. Three degrees have been outlined (Roth 1992). They emphasize benchmarks along segments of the continuum.

Nominal environmental literacy (at the lower end of being environmentally literate)

- Developing awareness and environmental sensitivity.

- Increased respect for nature and concern for how humans interact with it.

- Rudimentary knowledge of natural systems and how humans interact with them.

Functional environmental literacy (in the middle reaches of environmental literacy)

- Broader knowledge and increased understanding of human/environment interactions.

- Awareness and concern about negative human interactions with environment.

- Developed skills with which to analyze, reason, and evaluate environmental information.

- Ability to communicate conclusions and feelings about problems/issues to others.

- Willingness to act to resolve problems/issues of personal concern.

Operational environmental literacy (at the higher end of environmental literacy)

- Broad and deep knowledge of ecology and human-environment interactions.

- Routine evaluation of environmental impacts of human actions and consideration of their consequences.

- Active and regular gathering and evaluating of relevant information.

- Making decisions among alternatives, and advocating appropriate actions.

- Holding a strong sense of responsibility and personal investment for the environment.

- Acting at several societal levels, from local to global, as well as both personally and collectively.
- Living an ecologically sustainable lifestyle.

Note that these characteristics of environmental literacy apply to technologically developed societies, such as the United States. Many of the world's people do not live in such industrialized countries. Many people who live in pre-industrial or underdeveloped societies are in a qualitatively different situation. While it may be argued indigenous people "live in harmony" with the environment, this appears most true when they live without modern technology and with minimal contact with industrialized values. Our purpose here is not to assign blame for environmental degradation, however. Suffice it to say environmental degradation exists, and good communication practices can help to foster operational environmental literacy, through which sustainability by definition results.

Roth (1992) states that environmental literacy is the goal of all environmental education. Given you necessarily must communicate to educate, here's a direct corollary: *Environmental literacy is the goal of all environmental communication.*

HOW SCIENCE INFORMATION BECOMES RELIABLE

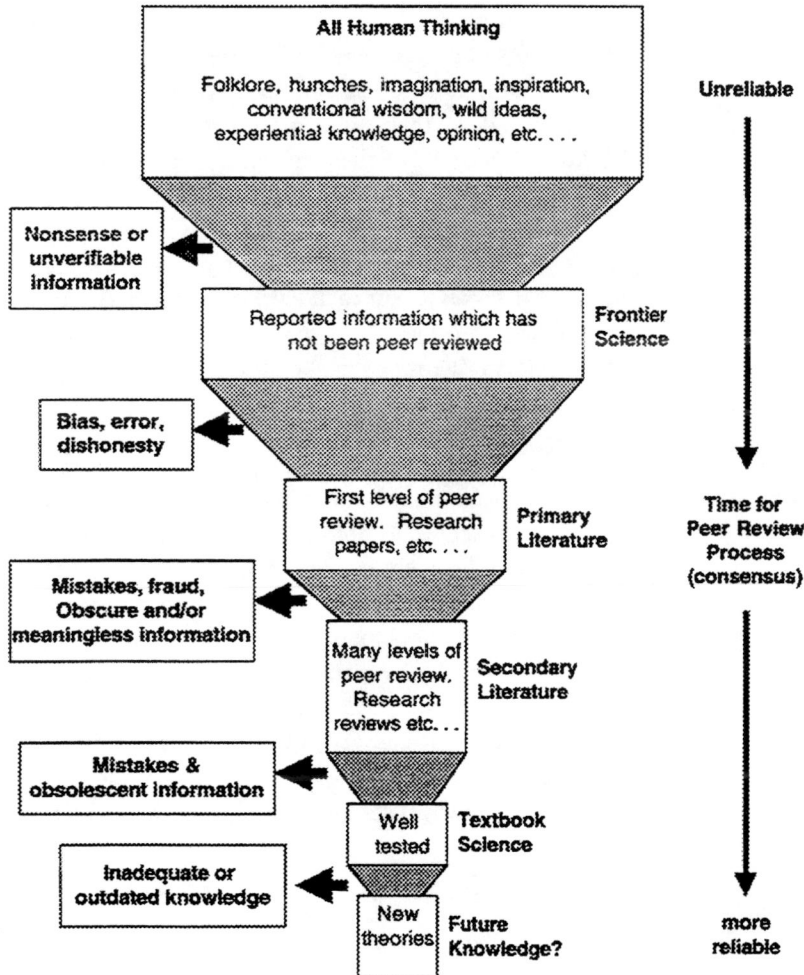

Figure 3   The Scientific Knowledge Filter

Being literate hinges on knowing the credible, valid, reliable, and trustworthy from that which is not. The "information superhighway" indicates incredible accessibility to vast amounts of information. But, the communicator needs to know good information from not-so-good information. This is especially true when it can appear on your computer screen with the same ease. Knowing how information becomes credible and acceptable within the scientific community can help.

Core to acceptance of research findings by scientists is the process of peer review. The schematic "knowledge filter" shown here winnows information by judging its merits based on the quantity and quality of review it has undergone by the community of science. Keep in mind that the knowledge filter strains out non-scientific knowledge. Information which cannot be verified and does not fit a consensus of the science community, after being reviewed by more experts, is removed. This is not meant to imply that all such knowledge is without merit. Indeed, emotions, spirituality, folklore, and intuition are vital components of human culture. They are not science, however.

At the top of the filter all information generated by all human endeavors is input. This input includes misinformation, misconceptions, folklore, and pseudo-sciences such as astrology, psychic phenomena, and fad dieting. Material from such sources does not fit the consensus of reality held by modern scientific thinking. In most instances they cannot be demonstrated by repeatable observation and measurement. This removes them from the realm of actual science. Still, these pursuits are not completely discounted as there are ways to convincingly validate them to many. There is enough circumstantial evidence from "believers" to continue their existence in the popular culture. When was the last time you read your horoscope?

Each subsequent step of the knowledge filter further screens ideas, hypotheses, contentions, and speculations. Credibility rises as the amount of unverifiable knowledge decreases. At each level more experts, who have extensive experience with similar information, review the newer material and judge it to fit existing ideas or declare it erroneous. We have applied labels to various steps along the way.

Frontier Science

After draining off nonsense and information that can never be verified, the knowledge filter continues to hold information that meets the most cursory criterion of seeming to be scientific. At this stage, this "scientific" material appears without the benefit of peer review. An example might be scientists who announce major findings through press conferences, in order to generate wide public interest in their work. Sometimes this is done as a tool to increase funding for further research, often focused on outlandish concepts. Many documents of frontier science are written and published by self-proclaimed experts. These books are generally not peer-reviewed. To the undiscerning reader, this information represents expertise. In reality, it should be judged with heightened skepticism.

Primary Literature

In the first stage of scientific peer review, a report is reviewed primarily for its face validity, guarantees of reliability, and contribution to a particular field. Face validity means that it holds together and seems to make sense during a close read-

validity means that it holds together and seems to make sense during a close reading. Reliability deals with stating the methods used and showing how the data were handled. To be reliable a study has to offer instructions for others to repeat the work. Because science is iterative, and builds on itself, a study most always connects with others. This connection may be to confirm or contradict other findings.

Primary literature includes scientific journals, research reports, academic books, and conference proceedings. Findings released at this filter level may, after closer inspection, be declared to have flaws in methodology, analysis, and interpretation. Fraud is also a possibility against which science is constantly vigilant. Consensus builds when data and results can be replicated and begin to make more sense when fitted into current theories. Material that is flawed or does not fit current theory is usually rejected, at least until more corroboration can be offered. It should be noted that many revolutionary ideas in science have been initially rejected, only to become future benchmarks. Revolutionary science, however, tends to be hotly debated before it is tentatively accepted by a minority in the science community, whence forth it then makes its way down the filter.

Secondary Literature

Generally scientific material that has been replicated and fits the orthodox views of the science community finds its way into the secondary literature. This material is composed of larger reviews and monographs written by experts in specific disciplines. They strive to compile many studies, fitting all the data into a larger picture. Much of this material may have first been published several years prior to being reviewed. Scientists who undertake reviews spend large amounts of time contemplating a body of work, judging the highly technical work rigorously.

Textbook Science

Scientific theory-building begins to take shape between the secondary literature and this level of the knowledge filter. After continuous acceptance during several years of replication and further study, valid and reliable ideas make their way into textbooks, compilations of the best works in particular fields. The next generation of investigators learns from these books. In such books, theories that form the knowledge structures of science are explained and purported to describe reality with parsimony.

As information passes through the knowledge filter, it gradually becomes more accurate as findings exhibiting bias and lack of rigor are filtered out. Note how time within the knowledge filter increases as information moves down each step of the way. Note also that consensus is based on the prevailing theories and models within each scientific discipline. Still, even highly developed concepts at the textbook level of the filter can be invalidated by newer thinking that develops better theories. This throwing out of once strongly-held theory occurs most often in rapidly developing fields, such as astronomy, quantum physics, and environmental science. In each of these cases, new instrumentation and techniques have recently uncovered previously unknown data.

An unfortunate attribute of the knowledge filter, the information most readily available to the non-scientific public is generally from nearer the top of the filter. This material is the most likely scientific information to catch the attention of the gatekeepers of the mass media, and these news outlets are where most adult

Americans learn about new scientific discoveries (Miller 1989). Contentiousness between journalists and scientists is something environmental communicators are apt to have to deal with on a regular basis (Friedman et al. 1986). A time and place where the contents of scientific textbooks make the news is almost assuredly far, far away.

THINKING CRITICALLY ABOUT SCIENTIFIC INFORMATION

Scientists collect data and then use it to generate information that then forms into new knowledge. The means through which they generate data, information and knowledge is the scientific method. Knowing how research is conducted allows a critical reader of scientific reports to ask the right questions in order to judge the worth of the study in their own search for environmental solutions. The following set of questions can guide a critical analysis of scientific information.

Does Their Argument Make Sense?

What is the crux of their argument? What are they claiming to have found? Many times a clear sounding statement such as "80% of the population of the United States are environmentalists because they have environmental attitudes" really tells us nothing useful. What is an environmental attitude? Is it one attitude or many specific attitudes that defines being environmental? Who was doing the defining? How was the 80% figure derived? We often have to make inferences and decisions based on insufficient evidence. Avoid "leaping to conclusions" by critically reviewing the information. Get to the core of their argument by closely scrutinizing the abstract and the discussion sections of the report.

Who Is the Source of the Information?

Examine who is making claims, and what their motivations can be inferred to be. Try to expose hidden agendas. This does not mean that either an environmental group's or an industry's positions are automatically biased beyond aceptance. Everyone can be considered a member of some special interest group. Call on your own reserve of knowledge, but also look for the source to offer alternative points of view or contrary evidence. Knowing what makes a source of information reliable helps to weed out the misinformation and propaganda (blatantly one-sided communiqués).

Are the "Facts" Placed in a Context of Accepted Knowledge?

Facts are always contextual, in that they must be understood within a bigger picture. Blanket statements need to be supported with information. Think of a statement claiming, "The number of cancer patients in America today is double what it was per capita when compared to 50 years ago." Before we conclude our modern pollution problems are responsible, we should first ask what background information is being left unmentioned? What ages are the cancer patients today compared to before? Is there an increase of older patients with cancer? What are the main causes of death today compared to before? Are third world nations developing heavy industry now showing the same trend? Modern medicine is improving so that we now have a population that is not dying from relatively minor diseases. Has this new trend been figured into the statement? Such questioning leads to a richer understanding of any subject.

How Was the Information Obtained?

We are bombarded with information all the time. It often sounds like "Chicken Little" is always predicting the sky is about to fall. When it doesn't, we are not surprised anymore. We are becoming desensitized to the myriad information we now receive. Questions such as "Says whom?" "Who paid?" and "What then?" get at the ways and means by which the information was gathered and then sent to you.

This line of inquiry is especially critical in the instances of "one study panic." A startling example of this happened in 1989, when many people in North America were scared with the report that Alar, a pesticide used on apples, was found in apple juice given to children. At the time most of the studies about Alar had shown it to be relatively harmless to humans, yet one report claimed that the Alar was detrimental to children. A major environmental group and a major national TV network championed a crusade against the Alar manufacturers and apple growers. The one contrary study was eventually dismissed by the U.S. Environmental Protection Agency, but not before a lot of damage to the apple growing industry. The moral: view findings in context with what is already known.

What Kind of Study Was Reported?

Most research studies can be grouped as either correlational or experimental. Did the study look at something that already existed and make conjectures about the outcome or did it control the variables and show the outcome as a cause-effect relationship?

Correlational research makes educated guesses that certain observations are linked in some logical way. For example, if a small town of people begin to have an increased number of cancer reports and at the same time the groundwater in the area is found to contain traces of the carcinogen benzene, then it might be assumed the two are linked. Since there is no tight control of all the variables that might have produced an outcome, further studies are usually necessary.

Experimental research controls all the factors to isolate the one or more identified variables (the cause) that force an outcome. Most drug and chemical studies are done in this way so that the result seen is attributed to the experimental compound alone. A test population, often cell cultures or non-human animals, is given different doses of a chemical and then observed to see when the chemical becomes toxic, and how the toxicity is manifested in observable symptoms. Later clinical studies in human volunteers are used to gauge reliability of drugs.

Were Measurements and Statistics Used Properly?

Statistics are complex mathematical calculations that indicate the significance of data. Significance here means statistically sound and not necessarily important. But, the misuse and erroneous reporting of statistics is cause for concern. If it is reported that the amount of a chemical found in a lake is significantly higher than it was a year ago, should we be concerned? Researchers refer to this as the "So what?" question. Finding data that reaches statistical significance is one thing, determining if it matters in the larger scheme of things is another.

Another important statistical consideration is the size of the study population. Though it is expensive to run tests on many subjects, small populations are more difficult to generalize from and introduce more uncertainty into outcomes.

THE ART OF ARGUMENTATION

As the last section demonstrated, asking insightful questions allows you to dissect the arguments contained in research reports. As an environmental communicator, astute members of your audience will be doing the same to your messages. Constructing solid arguments becomes indispensable to your success.

The art of argumentation is one of the most ancient skills attributed to learned persons. Rhetoric has a legacy that spans dozens of centuries. As we can see from the last section, asking the right questions is critical. Rhetoricians seek to build invincible arguments. Environmental communicators can borrow some rhetorical techniques to aid in the presentation of well-reasoned dispatches.

In stating and supporting a position—what we mean by arguments here, as opposed to anger-sparked tirades—your messages have to be able to withstand three essential questions that will be asked by critics:

1. What do you mean?
2. How do you know?
3. What was presupposed?

Bear these three deceptively simple questions in mind as you build your messages. Close and careful consideration of these questions will focus your attention on your presumptions, evidence, inferences, and the manner in which you combine them.

- **Presumptions:** A presumption may be defined as a statement of fact or belief for which no verification is required. These are technical points of agreement or beliefs accepted by participants as true. Usually any one objecting to a presumption has the "burden of proof" to show it untrue.

- **Evidence:** One selects bits and pieces of reality to support one's claims. These items are evidence. Evidence substantiates and attempts to verify what is being claimed. Empirical assertions have factual origins. Value assertions are drawn from aesthetic judgments. Policy assertions are based on ethical expectations. Each can be a source of evidence.

- **Inferences:** Inferences are derived from your evidence and are based on your presumptions. They are conclusions about what is believed to be correct. Reasoning is used to reach specific determinations. This reasoning can be deductive, where a specific point is reached from examining generalities; inductive, where a generalization is made based on specifics; or a combination of both. All inferences should be relevant and logical.

Arguments use a number of language devices to structure their content. Argumentative structure deals with the manner presumptions, evidence, and inferences are put together and presented to the audience. The most common language devices are analogy, metaphor, simile, and example.

- **Analogy:** To compare two things, which are alike in some respects, so as to explain a concept is to make an analogy. These extended explicit comparisons are common and useful to argument-makers. The familiar is used to explain the difficult to understand, highlighting similarities while downplaying or ignoring dissimilarities. An example might be that attempting to communicate to a hostile audience is like stepping into the lion's cage. It emphasizes that a lot of care and preparation needs to be in place before the communicator can expect the audience to be receptive.

- **Metaphors:** Metaphors equate one presumably familiar thing with another presumably less familiar one. Because they have great power in initial explanations, metaphors are abundant in everyday language. Yet, most metaphors exhibit vagueness and ambiguity under close examination. When a singular metaphor is used extensively, it tends to break down and lose its explanatory effect. An example might be that the communicator is molding an audience's opinion as a sculptor chisels a block of granite. The idea of crafting messages to develop a persuasive outcome seems fine, yet people are not inert pieces of rock. The rock is not influenced by anything but the sculptor. People are influenced all the time by factors outside the communicator's control, hence the metaphor breaks down under scrutiny.

- **Similes**: Similes compare a familiar thing with something less well-known, just as a metaphor. They, by definition, use the word "like" in making the comparison. This preposition serves to both highlight and qualify the comparison. An example is a communicator who argues that a small community trying to work with a mega-corporation over a local pollution problem is like David confronting Goliath. It emphasizes that the size difference is daunting, yet like David, the community can get the mega-corporation to listen to their grievances if the community squarely faces the problem and remains focused.

- **Examples**: When one thing is used to represent a group, an example is being made. Examples are inductive devices because a sample is used to show the validity of many. In explaining complex issues, examples are extremely useful, almost necessary, to staking a clear position. An example of this is when a previous oil tanker spill off a coastline is used to argue for establishment of regulations for all oil tankers in coastal waters.

The validity of any argumentation is only as good as the rigorousness of the presumptions, evidence, inferences, and language devices being used to construct it. Sound construction of an argument results from sound reasoning. Arguments are weakened by unsound components and faulty construction. Such weaknesses are referred to as logical fallacies. Argumentation experts (e.g. Makau 1990, Freely 1997) list fallacies as falling within the language, evidence, or reasoning of an argument.

Language

- *Ambiguity*—more than one interpretation can be applied to a premise.
- *Vagueness*—meaning is too inexact to contribute to the discourse.
- *Equivocation*—changing meaning during a discourse to make an argument seem more compelling than it is.
- *Obscuration*—hiding behind needless jargon, terminology, semantics, etc.

Evidence

- *Repeated assertion*—giving an argument over and over again in the hope that it will become more acceptable.
- *Non-representative instance*—using a poor example.
- *Insufficient instances*—not giving enough examples.
- *Invalid statistical measure*—using biased, atypical samples, and other inappropriate mathematics and statistics to bolster an argument.
- *Unreliable source*—using biased, non-credible, or inappropriate sources.

Reasoning

- *Straw argument*—using weak versions of opposing or alternative views.
- *Begging the question*—avoiding or circumventing the relevant issue.
- *Circularity*—using unsupported assertions or simply restating a claim using different wording.
- *Non sequitur*—stating an irrelevant claim that does not follow the argument's evidence.
- *Appeal to ignorance*—using untenable burden of proof to force acceptance of an assertion.
- *Appeal to popular prejudice*—relying on what most people think to force acceptance of a claim.
- *Appeal to tradition*—relying on conventional social practices to enforce the correctness of a position.
- *Ad hominem*—challenging the maker of the claim instead of the claim itself.
- *Oversimplication*—overlooking potentially relevant information in order to make the issue easier to understand.
- *Hasty generalization*—using small, biased, untypical samples as evidence for a broader group.
- *Post hoc*—erroneously trying to emphasize an alleged cause-effect relationship.
- *Faulty comparison*—drawing conclusions from unwarranted comparisons.

Environmental communicators will continually be constructing arguments. While there are many pitfalls to avoid, there are also many salient positions that need defending. Indeed, the concept of operational environmental literacy calls on practitioners to take firm stances in restoring, conserving, and sustaining healthy ecosystems. But, taking a position and making a stance need not be contentious. One can cooperate and still make a sound argument.

Cooperative argumentation focuses on reasoned interaction about a controversial issue, with the intention of helping participants make the best assessments and decisions possible under the given circumstances. Such a process usually leaves participants feeling better about the decisions made, since all involved have shared information and ideas. Such buy-in makes for longer lasting solutions to issues.

Counter to cooperation, competitive argumentation focuses primarily on winning a debate. It is often referred to as "combative interaction." This process alienates participants involved rather than clarifying a situation. Each participant concentrates on trying to either prove themselves right or to prove their opponents wrong.

Arguments can be competitive or cooperative, inclusive or combative. The participants in an argument most always have the power to decide which sort of interaction they wish to have. Allies and enemies can both be made through argumentation. Environmental communicators are wise to take a long-range view and strive for cooperation over competition, to make allies instead of enemies.

REFERENCES AND FURTHER READING

Arons, A. B. (1990). Achieving wider scientific literacy. *Daedalus* (Journal of the American Academy of Arts and Sciences), 112(2), 91–122.

Aylesworth, T. G., & Reagan, G. M. (1969). *Teaching for thinking*. Doubleday.

Bauer, H. H. (1994). *Scientific literacy and the myth of the scientific method*. University of Illinois Press.

Hamm, C. M. (1989). *Philosophical issues in education: An introduction*. The Falmer Press.

Freeley, A. J. (1997). *Argumentation and debate: Critical thinking for reasoned decision making*, (9th ed.). Wadsworth Publishing Company.

Friedman, S. M., Dunwoody, S., & Rodgers, C. L. (Eds.). (1986). *Scientists and journalists: Reporting science as news*. Washington, DC: American Association for the Advancement of Science.

Hacker, D. (1997). *Rules for writers: A brief handbook*, (4th ed.) pp. 313–334. Bedford Books.

Hill, B., & Leeman, R. W. (1996). *The art and practice of argumentation and debate*. Mayfield Publishing Company.

Hirsch, E. D., Jr., & Mulcahy, P. (Eds.). (1988*). Cultural literacy: What every american needs to know*. Vintage Books.

Janis, I. L., & Mann, L. (1979). *Decision making*. The Free Press.

Lakoff, G., & Johnson, M. (1980). *Metaphors we live by*. The University of Chicago Press.

Makau, J. M. (1990). *Reasoning and communication: Thinking critically about arguments*. Wadsworth Publishing Co.

Miller, J. D. (1983). Scientific literacy: A conceptual and empirical review. *Daedalus* (Journal of the American Academy of Arts and Sciences), 112(2): 29–48.

Miller, J. D. (1987). Scientific literacy in the United States. In *Communicating Science to the Public*, Evered, D. & O'Connor, M. (Eds). Wiley.

Miller, J. D. (1989). Scientific literacy. Paper presented at the *Annual Meeting of the American Association for the Advancement of Science* (San Francisco, January 17, 1989).

Orr, D. W. (1992). *Ecological literacy: Education and the transition to a postmodern world*. State University of New York Press.

Rieke, R. D., & Sillars, M. O. (1996). *Argumentation and critical decision making* (Longman Series in Rhetoric and Society), 4th ed. Addison-Wesley Publishing Company.

Roth, C. E. (1992). *Environmental literacy: Its roots, evolution, and direction in the 1990s.* Columbus, OH: ERIC Clearinghouse for Science, Mathematics, and Environmental Education.

Rybacki, K. C., & Rybacki, D. J. (1999). *Advocacy and opposition: An introduction to argumentation,* 4th ed. Allyn & Bacon.

Walton, D. N. (1989). *Informal logic: A handbook for critical argumentation.* Cambridge University Press.

INVESTIGATING ENVIRONMENTAL ISSUES

Environmental communicators find they spend a lot of time and effort considering solutions to environmental problems and investigating environmental issues. Issue analysis was central to the making of the optimally environmentally literate person described in the last chapter. An ability to understand the roots of environmental problems and issues better equips one to effectively communicate about how to resolve them. First, we need to distinguish between problems and issues.

Problems are smaller units of controversy, whereas issues most often are larger societal disputes. Though this distinction can be fuzzy, problems tend to conglomerate into issues. A problem can be thought of as singular and more likely to involve an us-against-them position of participants. They can be extremely difficult to solve, because of this bipolar stance between the players. Often the two sides are forced into a winner-take-all situation where the losers get nothing. Not surprisingly, problems are either actually or potentially combative and foster antagonistic attitudes. For example, environmental problems such as whether to log a forest, strip-mine an area, or dam a river hinge on singular decisions. Because of this either-or nature, problems lend themselves to rapid understanding. Still, stakes can be high and outcomes costly to the loser.

Issues, on the other hand, are much more difficult to comprehend. They have many facets and aspects to be managed if they are to be resolved. Because issues are constituted from collections of problems, they can be termed multidimensional. Examples of issues include global climate change, air and water pollution, land use management, endangered species, and human population growth. There are many, many others. Notice how the labels applied to issues give them a luster of scientific detachment and do not really describe how people feel about them.

Issues do not usually hinge on singular answers to questions. Instead, they tend to involve the answering of chains of questions. Each answer is commonly greeted by the arising of new questions. As a series of answers are reached, they drive management of the issue. Whereas problems dissipate after one decision, issues require strings of decisions to mitigate. Resolution of issues takes a long time and sometimes is never fully reached.

Consider some environmental issue-linked questions: How should wilderness be managed? Should human population be managed? What constitutes acceptable levels of risk for industrial chemicals? How do we resolve conflicting interests for land uses? How can tropical rainforests be preserved? What are the most prudent uses of pesticides and herbicides in farming? Answering such questions taps directly into beliefs and values upon which positions are established.

Issues hold meaningful importance to individuals, groups, and communities. They have consequences, either real or imagined. Although hugely relevant in confronting issues, importance is extremely difficult to measure and not much easier to characterize. Describing what an issue means to a party comes from understanding the party's beliefs and values that underlay their position on the issue. And, there is always more than one position to consider.

Because individuals, groups, and communities start dealing with any issue from different places, different perspectives are inevitable. Commonly, there are several points of view. Over time, however, multiple points of view usually devolve into polarized sides. Each of these two sides supports an often-oversimplified opposing theme, which remain after alternate minor points of view have been subsumed. Have you ever heard an issue summed up as "jobs vs. the environment"?

Attempting to resolve issues tends to cut to the core of a person's belief and value systems and so must be treated with great deference. The very act of challenging a position on an issue can be tantamount to challenging the person directly. *When beliefs and values are threatened, the result is outrage and fear.* Outrage and fear further compound issues by making it more difficult for those with differing points of view to communicate. This fear-based type of noise can be severely damaging to any communication system. Systems can utterly break down when infected with fear. Outrage can also degrade into physical violence. Fist fights, terrorist acts, and even wars have come about when values and beliefs about environmental issues have been offended. Clear and cogent environmental communication early in an issue's emergence provides some insurance against such unfortunate outcomes.

COMPONENTS OF ISSUES

- **Problem**—A situation occurring in which someone or something is at risk. The problem is agreed upon, but the solution or process is usually in dispute.

- **Issue**—A multidimensional situation about a complex problem or situation where different values and beliefs are held by various players/stakeholders. Usually there is little consensus on how to address the priorities of the various dimensions within the issue. Questions that begin to elucidate an issue might be: What is at risk? Concepts will vary on the differing perspectives of the stakeholders. Is this same object given by all the stakeholders? If not, how do they differ? What is each stakeholder or group really referring to in the issue? It is essential to define the situation clearly for all positions. Are the needs real or just perceived needs?

- **Players/Stakeholders**—Individuals, groups, or organizations that have a vested interest in the issue and its outcome. Who are the individuals, groups, or organizations that have a role in the issue? Are there players who might be involved, who are not stakeholders, but can complicate the group dynamics when trying to form a solution? What kinds of vested interests exist that might detract from the defined issue?

- **Positions**—The stance or postures that the various players adopt concerning the issue. What is the situation (problem or its solution) for which differing values and beliefs exist for the various players. Do you understand why the various players hold the beliefs and values that they do?! Are the players defending their positions based on valid information? Is the decision being made by all players an informed one based on reasoned thinking, or is re-clarification needed which requires additional information? Are all alternatives given critical review, or is it possible that a groupthink situation exists? (See Chapter 10 for more on groupthink).

- **Solutions**—Alternative strategies and ideas employed to find consensus and resolution for an issue (Ramsey et al 1989). Is a compromise between the players possible? Is a consensus possible? Do all the alternatives and solutions proposed address the defined issue and not just vested interests? Who will be affected by the short and long term consequences of any solution? What are the costs and benefits to all players associated with any solutions? Is the issue still easily resolvable? If not, then dispute (conflict) resolution may need to be implemented (conflict management is covered in Chapter 17).

Box 1: Value Descriptors

Values undergird all messages. They are tightly held by all within a communication system. Any lasting resolution of an issue will have to reveal and address the values of each group of stakeholders (Hungerford et al 1998). Here are types of values within communication systems:

Aesthetic—It is often said "that beauty is in the eye of the beholder." What looks like a plain piece of scrubby grassland and briar patches to one person, may appear to another like a small desirable piece of wild area inhabited by numerous small forms of wildlife. A carpet of dandelions may be beautiful to one person and a scourge of weeds to another.

Cultural—Many communities may have conservative approaches to doing something, because it suits them to do it that way and it has "always been done that way." Challenges to established methods of doing something usually meet with resistance.

Ecological—People tend to resist even incremental changes in an area that is unique to them. Putting a road across a wooded hillside may create a barrier to wildlife.

Economic—How might any actions affect the economics of an issue? Who stands to gain and who stands to potentially lose? In the case of loggers losing their ability to harvest trees on national forest land, how are they to continue to make a living?

Educational—What unique learning experiences can be gained from the issue and also from the process of managing the issue?

Egocentric—Refers to a focus on self-satisfaction and personal fulfillment; a "me" oriented value.

Legal—People may like or dislike legal values depending on how it affects them personally. If a law prevents hunters from hunting out of season, they may challenge it. Yet another group wishing to stop hunting may support and even campaign to extend the non-hunting season.

Recreational—How people spend their time is important to them. If an issue affects their recreational activity or takes too much time away from it, then resistance will be encountered.

Spiritual/Religious—It may correspond to an organized religion experiencing the connection, or to the more esoteric internal connection with something beyond the human world.

Social—People come together for a variety of shared reasons such as empathy, feelings, or status. Saving the rain forests of South America has drawn many people to join organizations to stop the clear-cutting of these forests. In these groups these people feel camaraderie and a feeling that they are doing something useful together.

How Issues Arise

Issues do not spontaneously arise in modern societies. They require champions who seek to amplify a set of concerns and highlight a group of problems. Issues arise because a group, or a particularly vocal individual, decides to "push" their agenda into wider recognition. Champions have to gain attention, build support, and drive expansion of concern. All these functions are based in communicating well.

Protecting the canyon lands and red-rock country of the Colorado Plateau as federally designated wilderness is the mission of the Southern Utah Wilderness Alliance (SUWA). The group formed in 1983 and seized on a deceivingly simple goal of 5.7 million acres of newly declared wilderness in Utah (Shapiro 1998). Using "5.7" as a rallying cry, the group generated support throughout the United States, growing from zero to 20,000 donors in less than a decade. More importantly, the group was able, through wise and prudent communication strategies, to take an administrative problem and transform it into a broad public issue. The federal Bureau of Land Management was ordered in 1976 to study its Western land holdings and make recommendations to Congress for wilderness designation. In Utah, the BLM found 1.9 million acres worthy of this protection. Environmentalists disagreed and SUWA was established. The group's organizers decided to forego political lobbying and, instead, build grassroots support. They drew on Utah's sole metropolitan area, the Wasatch Front where two-thirds of Utahns live and then moved their campaign to the East and West coasts. The messages used by SUWA tapped into the beauty of the "intricate canyons, arches, and vast expanses of slickrock" in Utah (Shapiro 1998, p. 265) as well as the learned legitimacy of scientists familiar with the ecology of this country. Attention also had to be given to the fears of residents of southern Utah, who felt they would be removed from their land. Throughout, SUWA successfully built a powerful coalition, by skillfully blending mass-mediated and interpersonal messages, taking alternately cooperative and confrontational stances, and creating an environmental issue that would not exist without their championing it.

SUWA and their beloved Utah wilderness illustrate numerous dimensions of issues. They are nurtured into wider consciousness. They call for a mix of messages to further the cause. They require attention on many levels—regulatory disputes, conflict management, public relations, legal repercussions, and health aspects. An issue cannot arise without confrontation and conflict, though these can be diminished. Opponents who are not adequately communicated with can erupt out of fear and deep disagreement. A complex situation can get unduly complicated. Managing messages is key to handling issues. And, understanding is a prerequisite to crafting successful messages.

Dissecting Issues

Analyzing issues is a lot like dissecting a specimen in a laboratory. You have to get beyond the surface to see what is going on inside. Different parts serve different functions. Perhaps the most difficult item to deal with is finding the core argument held by each side. Opposing groups many times have tremendous difficulty grasping what others are actually contending. Clarity of rhetoric is rare in environmental issues. Groups posture and position themselves. These stances shift constantly. Getting the upper hand in a battle of argumentation seems to require continual rearranging positions and an arsenal of rhetorical strategies. It is common to see two groups in a heated debate getting exasperated with each other because neither seems willing to see the other's viewpoint. This is often because the groups are locked into a confrontation over mis-aligned problems. The groups are not arguing about the same viewpoint at the same time. Though each problem may be set within a general framework of an issue, they still may not be parallel. Arguing about them as if they are can only increase frustration.

The famous case of logging in the Pacific Northwest versus protecting the spotted owl clearly shows this situation. One group is arguing about jobs and lifestyles while the other is arguing about ecosystems for an endangered species. While both viewpoints are valid, it is necessary to realize that each viewpoint must be resolved independently.

Communicators need to be ever vigilant about the multidimensional nature of issues. Confusion and clarity can both result when discussing issues. As groups declare their positions and discredit those of their opponents, watch for these characteristics within their perspectives:

- **Biased**—over-emphasizing a selective viewpoint demonstrates bias. One definition of "propaganda" is material that espouses one and only one position. Even though information put forth by issue-interest groups is usually agenda driven and meant to persuade you to adopt their point of view, it is essential to issue resolution to understand all perspectives in an issue if resolution is ever to be achieved.

- **Simplified**—sometimes so much information about an issue is left out of an explanation that the message is misleading. Certainly, making a position easy-to-understand is laudable. Leaving out relevant details is not, however. Over-simplification is often a problem in risk communication scenarios, where crucial scientific details are omitted by message-makers who see shortcomings in their audience's scientific background instead of their own explanatory skills.

- **Personalized**—when individual human dimensions of a wider environmental problem are highlighted, the issue is given a face. Such anecdotal evidence is powerful and persuasive. It tugs at your heartstrings. When used exclusively, though, personalization can cloud larger and more substantial aspects of an issue.

- **Sensationalized/Glamorized**—similar to being personalized, an issue is sensationalized when the focus is on a single town, natural area, industry, or other entity. While the impact on the entity in the spotlight may be great, the connection to the wider society can be lost. When glamorized, an issue is championed by a celebrity figure. This may provide a lot of publicity, but the issue often plays second fiddle to the celebrity's persona.

People within groups sometimes erroneously think that because they agree on an overall solution to an issue, their positions on answers to problems within the issue must be similar as well. Members of a group are more likely not to be in agreement on such details, even if they thought so on first brush. Values on which reactions to problems are based differ by individual. So, even within groups compromise will be necessary.

A vivid example of this occurred in the Sierra Club in 1998. At question was whether the Sierra Club should take an official policy for limiting immigration into the United States. Following the most extensive and vociferous internal debate in the club's history, members voted to maintain a neutral stance on this issue (Cone 1998). The campaign, which ended with a 60-40% vote, featured mass mailings, pithy sound-bites, and accusations of racism. Even as this campaign created factions within the Sierra Club, there was no question that the nation's most powerful environmental group was united in its view that human over-population must be stemmed to reduce environmental degradation.

FRAMING AND FRAMING ANEW

A concept useful in describing and analyzing group positions, framing refers to the perspective from which an issue is viewed. To frame is to select certain dimensions of the issue at hand and grant them more salience than others (Entman 1993). Messages are framed by the way their component bits of information are included, connected, and arranged. Frames define issues, diagnose causes of problems, and suggest answers and solutions. Conversely, framing always involves exclusion of portions of an issue. Frames are based on values that drive the selection process and the measuring of salience. The term "frame" is itself a metaphor, helpful in generating an understanding of an intangible mental process. Like a border on a painting or photograph, a mental frame sets off the subject from the rest of the field of view. Frames put boundaries on reality, tend to be self-reinforcing, and are difficult—but not impossible—to overcome (Entman 1993).

The initial task in revealing a frame is to find out who is controlling the composition and release of information. Historically and currently, information providers have tremendous power, and power is the prime factor in pushing one's own agenda over somebody else's. Therefore, information and the way it is framed exerts control over the audience. Audiences have been shown to take cues from frames, learning what issues to pay attention to and what connections exist between problems within an issue.

Environmental communicators should look for indications of frames as they gather information about issues. This need not be a strictly academic exercise, as long as one considers that frames have power to impose meaning. If one can find signs of the frame in place, one can begin to construct alternative frames. Changing frames is possible and useful. Done effectively, new frames can build dialogue and result in more harmonious relationships among groups (Ryan 1991).

Redefinition of frame is a typical tactic used in environmental communication. In using this re-framing tactic, you change the original point of contention in order to obscure an opponent's argument, gain support from a broader audience, or align your own message with some value of the target audience.

In establishing a regional outreach program, the Idaho Museum of Natural History conducted audience analysis to better understand the knowledge, interests, and attitudes of the people of Idaho (Sommer 1999). This front-end evaluation showed the museum that the traveling exhibits, educational videos, and traveling trunks they wished to develop and disseminate needed to reflect the values of rural Westerners if they were to facilitate learning about biodiversity. They decided to do this by using language that was common to and acceptable by the audience, even if it was unconventional for natural history museum staff. The outreach program, called the Natural Heritage Project, is built on a teaching metaphor, "the economy of nature." This metaphor is used instead of "ecosystem" which was found to be potentially threatening. Likewise, instead of "environment" and "evolution" the Natural Heritage Project uses "the great outdoors" and "natural selection." The political baggage carried by the conventional terms is, thus, avoided. The reworking of language amounts to an aligning of the frame embedded in the museum's educational messages with the values of its intended audience. Curators and educators wisely adopted an alternative frame that would facilitate learning by their audience, rather than generate noise in the form of confusion and anger.

A pair of cautions are in order. First, note that lapsing into propaganda is easier than it should be when constructing an alternative frame. Propaganda is not always the telling of lies. It can also be presenting only selected truths to support your position, while completely neglecting other viewpoints. Second, when issues are redefined it is common to retain presumptions from before. These should be examined and probably changed as well, however. This keeps groups from assuming they are still arguing the same points as before and getting into the uncomfortable and unproductive situation of "talking past each other."

Properly understanding issues is the first, vital step in attempting to manage them. This chapter has dealt with the organization of information as it pertains to revealing the complexities of environmental issues and the values of the groups staking and defending positions on such issues. Understanding an issue clearly makes the prospect of getting all participants talking about the same topic at the same time achievable.

Box 2: History of an Issue: Ohio Beverage Container Deposit Legislation (BCDL)

Over the last two decades in Ohio, there has been a drive to pass BCDL. The proponents, a group calling themselves Citizens for the Environment, wanted a deposit placed on beverage containers to prevent them from being discarded after use. Such legislation would provide a financial incentive to bottling manufacturers to pick up containers for recycling from consumers. Opponents of BCDL do not want to place such a burden on the industry.

In 1994, the beverage industry redefined the issue completely. They offered taxpayers an alternative frame by declaring the penny-per-bottle tax was a tax on food. Food taxes are both unpopular and unconstitutional in the state of Ohio. The bottling industry placed a referendum on the November 1994 ballot and launched a campaign that convinced Ohioans to repeal this tax. The bottling industry was able to concentrate the public focus on food tax and not litter reduction and recycling. Voters repealed the tax, thereby taking $64 million out of the state's annual budget and giving it back to the bottling industry.

REFERENCES AND FURTHER READING

Bryant, B. (1995). *Environmental justice: Issues, policies, and solutions.* Island Press.

Cone, M. (1998) Sierra Club to remain neutral on immigration. *Los Angeles Times* (April 26, 1998).

Entman, R. M. (1993). Framing: Toward clarification of a fractured paradigm. *Journal of Communication,* 43(4), 51–58.

Goldfarb, T. D. (1999). *Taking sides: Clashing views on controversial environmental issues,* (8th ed.). McGraw Hill College Division.

Hungerford, L., Peyton, Ramsey & Volk. (1988*). Investigating and evaluating environmental issues and actions: Skill development modules.* Stipes Publishing Co.

McConnell, R. L., & Abel, D. C. (1998*). Environmental issues: Measuring, analyzing, and evaluating*. Prentice Hall.

Ramsey, J. M., Hungerford, H. R., & Volk, T. (1989). A technique for analyzing environmental issues. *Journal of Environmental Education,* 21(1), Fall 89, pp. 26–30.

Ryan, C. (1991). *Prime time activism: Media strategies for grassroots organizing*. South End Press.

Shapiro, D. W. (1998). Lessons from a successful grassroots environmental campaign: The Southern Utah Wilderness Alliance. In S. L. Senecah (Ed.), *Proceedings of the Fourth Biennial Conference on Communication and Environment* (pp. 265–270). Syracuse, NY: State University of New York College of Environmental Science and Forestry.

Sommer, S. A. (1999). *Opening doors of communication*. Presentation at the Annual Conference of the American Association of Museums, Cleveland, OH (April 1999).

West, B., Sandman, P. M., & Greenberg, M. (1995). *The reporter's environmental handbook*. Rutgers University Press.

SECTION 2:
COMMUNICATION PLANNING

5

PLANNING ENVIRONMENTAL COMMUNICATIONS

Messages are most likely to be effective when they are part and parcel of planned campaigns. The process of communications planning formulates your campaign's goals and objectives, analyzes your intended audience, marshals available resources, and sets a schedule for its implementation. A plan's purpose is to harness and focus the power of the resulting communications system, to make it efficient and effective. This chapter presents an outline for writing plans for environmental communications campaigns.

As with any planning, it is handy to remember President Dwight D. Eisenhower's 1957 admonition: "Planning is everything; the plan is nothing." Planning is a methodical approach to a process. Once completed, the plan may need to be revised during implementation. As conditions change, so should your plan. Though it is important to be as complete and detailed as possible, flexibility in the execution of your plan will be advised if you find audiences reacting differently than anticipated. Flexible plans contain alternative actions for the most likely contingencies, and are instilled with the realization that the unpredictable can, from time to time, happen.

A planning process is diagrammed below. Each of its parts are detailed in the remainder of this chapter. Further information on each part can be found in later chapters as well. Keep in mind that these parts are constituents of a process. After detailing the process, we present a generic format for the plan document. Even though what we lay out here looks like a linear procedure, in reality planning is iterative and often roundabout.

A PROCESS FOR PLANNING CAMPAIGNS

The schematic below focuses on the steps of planning and executing a campaign. Each step needs to be considered carefully.

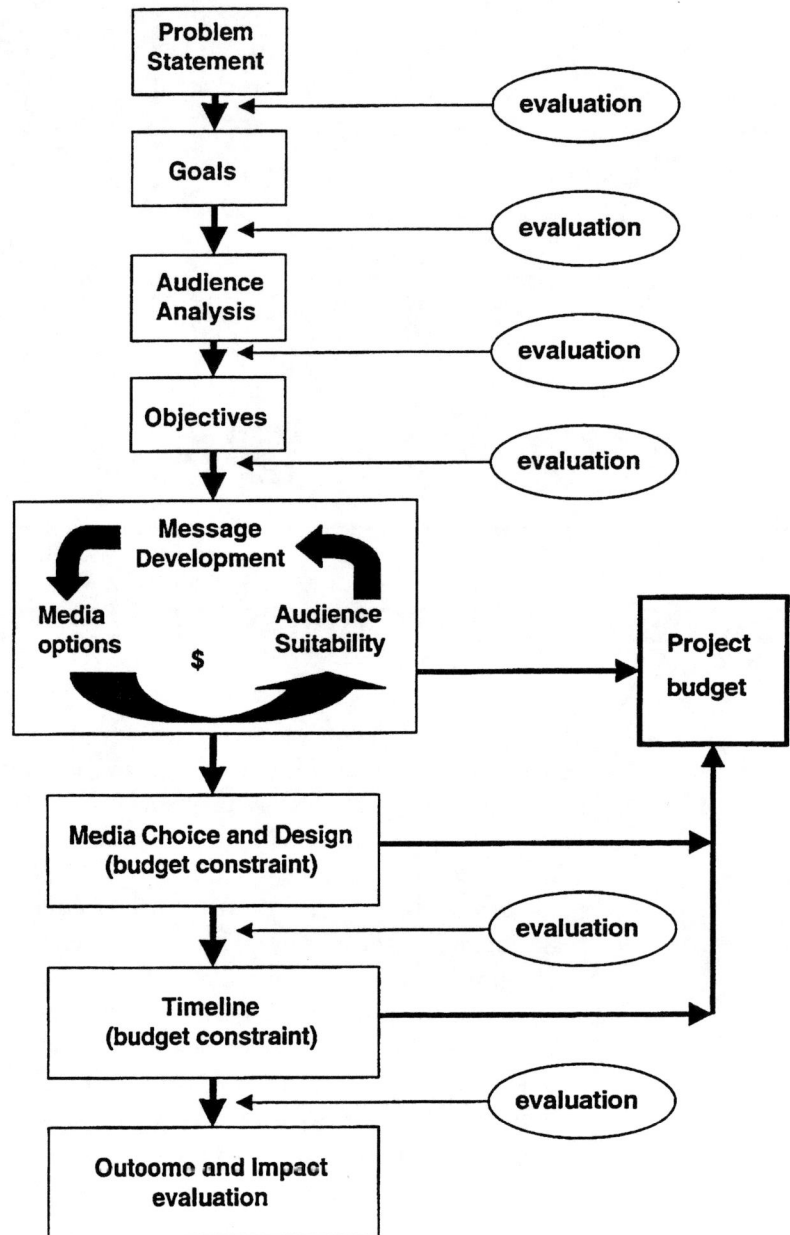

Figure 5   Communication Planning Model

Problem Statement

Communication campaigns are launched in response to specific issues, problems and needs. Remember from the last chapter that issues are multidimensional. A planning process begins with a thorough analysis of the subject to be addressed. This analysis is based on the principles discussed in the last chapter. After careful consideration and exploration of the aspects of the issue, a concise issue statement should be developed. Your problem statement guides the rest of the planning process. Research has shown that how environmental problems are defined affects how solutions to them are developed (Clark 1997). So, in order to understand the issue from other points of view, outside contributions should be solicited when developing an issue statement.

After drafting a satisfactory issue statement, it should undergo evaluation, to determine whether the statement adequately covers your position on the issue and is understandable by the intended audience. For example, you might convene a focus group to check for consensus on the adequacy of the issue statement. You may even have to do a needs analysis to identify just what is required. Regardless of the method you use, care should be taken to consider a range of opinions. You'll find more on evaluation in Chapter 7.

Goals

Once you are satisfied with your issue statement, goals for the communication campaign should be developed. Goals represent the ultimate aim of the project, and are best when they reflect long-term, lasting outcomes. In general, goals point to a resolution of the problems identified by the issue statement. Inclusive, open wording of goal statements allows more creativity in developing your campaign; specificity should be reserved for objectives as discussed below. Goals can be thought of as *qualitative statements of desired end-states.*

Audience Analysis

At this stage in the process, communication planners should identify and gather information about the specific audiences of the communication project. An audience can be conceptualized as a group of people who can be reasonably expected to react in similar ways to a message. They are those to whom your messages will be delivered (recall the receiver in the communications model from Chapter 2).

Audience analysis involves repeated bouts of defining and information gathering; it is iterative. As planners learn more about potential audiences, they refine their definitions of them. Well-done audience analysis makes other planning activities easier. So, it is important not to rush through this critical step in the planning process. Audience analysis is discussed in detail in our next chapter.

Objectives

Objectives are specific, measurable outcomes to be achieved by your messages. Objectives should be expressed in a way that makes it clear when they have been achieved. In addition, objectives should be stated as impacts that are outside of the direct control of the communicator, not as actions to be taken. This makes them distinct from the chronological steps of the timeline, directly under the communicator's control. Objectives are usually expressed in terms of a target audience taking some measurable action by some date. Some examples of objectives are:

- 70% of students at Smith Elementary School will recycle their soda cans by June.
- All park visitors will be aware of new trail regulations after January 1, 2003.

Objectives that are too vague for measurement, involve poorly defined audiences, or are expressed in terms of activities within the control of the communicator are not supportive of good planning. Examples of *poorly conceived* objectives include:

- All members of the state chapter will be more environmental ("more environmental" is too vaguely worded for measurement).
- The general public will support the timber sale scheduled for June 30 (the audience "general public" is too poorly defined for measurement).
- The communication team will issue a news release by December 1 (this is completely within the control of the team; it is not based on the audience).

The definition of objectives is a critical step in communication planning. Adoption-diffusion theory and locus of control, important concepts for framing objectives, are discussed in Chapter 6. In short, objectives can be thought of as *quantitative statements of the campaign's benchmarks and milestones.*

Message Development/Media Options/ Audience Suitability

After audiences are identified and objectives set, communication planners must formulate a message (or messages) to achieve their objectives. Simultaneously, media options, otherwise known as communication channels (see Chapter 8), should be researched for their abilities to deliver your messages. Each medium and message combination should be considered with your audience in mind. For example, simple messages designed to raise awareness for large publics might be planned for use with television or large-circulation newspapers, as long as you can confirm that the target audience use these particular channels. Narrower and more diffuse audiences require more selective media choices. Like so many cases in environmental communications, media placement is an iterative process, where continual review of your messages, media, and audience is imperative.

In order to move past this step of the planning process, a campaign cost target should be established. Some media options may be outside of the financial constraints on the project. For example, paid television advertisements will probably be beyond the monetary limits of small nonprofit groups. Though a detailed budget is not needed at this stage, simple aggregate spending limits are required for the plan to develop.

Media Choice and Design

After media options are identified, specific choices must be made. Given overall budget constraints, communication planners must decide which media options most likely will be effective in delivering the message and achieving the plan's objectives. Preliminary design should occur at this point, so that implementation planning and budgeting can occur. For example, if magazine advertisements are to be used, preliminary design work would include advertisement size and frequency, with mock layouts needed to continue the planning process.

Timeline

When media have been chosen and preliminary designing completed, communication planners must schedule the remaining steps needed to implement the plan. These steps should include specific actions to be undertaken, dates that the steps will be started and completed, and an assignment of responsibility for completing each step. This phase is completed when the communicator has developed an overall project timeline. A well-formulated timeline allows the communicator to carefully monitor a project's progress and to quickly identify bottlenecks, should they arise.

Formative Evaluation

Formative evaluation takes place throughout the planning process, during the formation of your campaign. It is used at the outset of a project to gather information and then, as the plan writing progresses, to assess whether a project is meeting its objectives. In communication planning, formative evaluation is an important tool to verify assumptions and decisions used during the process. Evaluation makes it possible for errors in assumptions or poor decisions to be corrected during the planning process.

Throughout every step of the planning process, formative evaluation can be used to confirm the decisions being made. For example, after the issue statement is formulated, planners can use formative evaluation to be sure that it accurately reflects the problem as seen by the sponsor. Likewise, goals can be verified using consensus techniques discussed in Chapter 7. Formative evaluation can be used to confirm that audiences are correctly identified, and to test messages and media with specific audiences. Rarely are complicated and expensive communication projects launched without first testing various components for effectiveness. Evaluation strategies are covered in Chapter 7.

Summative Evaluation

Communication planning is incomplete and risky without evaluation built-in throughout a campaign. Environmental communication plans should always include a description of how the project will be evaluated. Specifically, planners should identify specific means to be used to analyze whether the project has successfully met all of its objectives. All too often, communication efforts simply assume that the objectives will be met and never actually measure successes or failures. Most funding entities now insist that evaluation be provided, however. This creates accountability for resources being used in the communication campaign. While well-conceived objectives will ease the evaluation effort, specific evaluation techniques should be identified well before you need to call on them to see if objectives have been met. (Project evaluation is discussed in Chapter 7.)

Project Budget

Sound financial management is part of all good planning processes. The budget should include the cost of all materials, labor, purchased services, and overhead expenses for the project. One method for developing a budget is to review the project timeline and identify the costs of every step on it. Many planners underestimate the real costs of communication projects, and are later forced to prematurely end or scale back their effort, and thus not achieve their objectives. Detailed, accurate budgets greatly increase the professionalism of communication plans.

An Outline for Writing a Communication Plan

Each communication plan must take its own form. That said, the following format suggestions can be effectively used for compiling a communication plan. Such a plan should be professionally presented as a bound report with a cover page, table of contents, and identified sections. Appendices should be added to include additional material as necessary. As with all written communication, clarity, completeness, and brevity are important.

- **Abstract.** A one-page summary of the plan should be included at the beginning of the document. See Chapter 9 for guidelines for preparing an abstract.

- **Introduction/background.** The issue statement and a concise history of the subject being addressed, at the least, should be included in the plan. Planners cannot assume that all users of the plan understand the issues being addressed.

- **Goals and objectives.** The campaign's goals and objectives, as defined in the planning process, are included at this point. Remember that objectives should be specific, measurable, and stated as effects on the audience.

- **Target audience.** This section describes the target audience(s) of the communication project, including specific information about them and inferences made in designing and delivering messages to them.

- **Implementation.** This section should present the message(s), media choices with justifications, and the details of the plan implementation. Included in this section are the project timeline and budget considerations. Contingencies for revamping the campaign during implementation may also be included here. This section should be as detailed as possible, so that an individual who did not assist in preparation of the plan could still use it for implementation.

- **Evaluation.** This section should tie closely to the objectives. Evaluation methods should be clearly and succinctly described, along with criteria for determining the success or failure of the project. Evaluations may include all of the types defined in Chapter 7, or just one—it will depend on what the plan and objectives warrant.

- **Budget.** This section should itemize and detail all the costs of the plan. This helps the readers to see all the expenses involved and assists decision-making when adjustments are needed.

Planning is a prerequisite of success. It helps identify what needs to be done and the best ways to achieve the required outcome. Most of all, a good plan uses resources wisely and efficiently, and helps in avoiding unexpected problems.

REFERENCES AND FURTHER READING

Allen, J. (2000). *Event planning: The ultimate guide to successful meetings, corporate events, fundraising galas, conferences, conventions, and other special functions.* John Wiley & Sons.

Barban, A. M., Cristol, S. M., & Kopec, F. J. (1994*). Essentials of media planning: A marketing viewpoint.* NTC Publishing Group.

Bryson, J. M. (1995). *Strategic planning for public and nonprofit organizations: A guide to strengthening and sustaining organizational achievement.* Jossey-Bass Publishers.

Cassidy, A. (1998). *Practical guide to information systems strategic planning.* Saint Lucie Press.

Clark, T. W. (1997). *Averting extinction: Reconstructing endangered species recovery.* Yale University Press.

Donnelly, W. J. (1996). *Planning media: Strategy and imagination.* Prentice Hall.

Goodstein, L., Nolan, T., & Pfeiffer, J. W. (1993). *Applied strategic planning: A comprehensive guide.* McGraw-Hill.

Katz, H. E., & Knudsen, A. (1995). *The media handbook: A complete guide to advertising, media selection, planning, research & budgeting.* NTC Business Books.

ANALYZING YOUR AUDIENCE

Selling a product requires a thorough knowledge of the potential buyers. Marketers acquire and apply such knowledge about consumers. Likewise, environmental communicators anticipate particular reactions to their messages by audiences. Sending messages that produce the desired effects requires a thorough knowledge of the groups with whom you will communicate. Audience analysis comes early in the communication planning process for many reasons, because appreciating and catering to the attitudes and opinions of the groups your messages reach is crucial to the success of a campaign.

A target audience is any group for which a message is specifically developed and intentionally focused. An intended audience is one that the communicator expects to react to a message. This is not everyone who might see the message. The more you know about your intended audience, the more likely the message is to be received and acted in accordance with your campaign's goals. By action, we mean anything from becoming aware of a situation to permanently modifying a behavior within the audience. As you might suspect, making an audience aware of something is a lot easier than changing their behaviors. In this chapter we will deal with several aspects of analyzing audiences to promote an understanding of why people may or may not respond to your messages.

Learning about those who are targeted by your messages concentrates your campaign, keeping you from trying to reach too broad an audience. Audiences can be assessed by answering such questions as:

- What is the message? Why send it?
- Who is (are) the audience(s) that needs to be addressed?
- Why communicate with them? What makes them important to the success of your campaign?
- What makes them special and how can you customize your messages to meet their needs?

- What do you want them to do with this information? What kind of reaction do you expect?
- How will you know that they understood your message?
- What was their reaction to your message?

Keep in mind there is no such thing as a monolithic group called "the general public." Even though this term is heard often, it is far too imprecise to serve a useful purpose to environmental communicators. Using this term shows that audience analysis is missing or inadequate. Even if you have the aspirations and budget to try to reach millions of people, they still are not everyone. No message yet has reached all six billion humans on earth. Those with the widest reach— logos of such mega-corporations as Coca-Cola, Nike, and General Electric—may be seen by most, but nowhere near, all of us.

For environmental messages, there are many possible audiences. Each may be classified into a common group based on their shared characteristics, interests, and demographics. These are only indicators of your audience, however. Moreover, a group that accurately constitutes an audience for one type of message may not hold together for a different message. Audiences are fluid and ever-changing. Think of them as groups of people using an elevator: every trip up and down has a different makeup. Thorough research into your target audience is essential to maximize success with your communication efforts.

Recall the communication model. There should be a need to communicate that requires you to send a message to specific receivers. To reach them you'll have to overcome noise that interferes with the system. Encoding the message correctly will help reduce noise so that your target audience receives and decodes your messages as intended. Correctly analyzing your audiences is one major way of overcoming barriers to communication and ensuring that your message gets through.

Audience analysis usually involves talking to or surveying a select group of people that are similar to the wider audience you wish to reach. Such investigations are part of the planning process that goes hand-in-hand with having clear goals supported by measurable objectives and well thought-out evaluations. If you cannot get firsthand information about your audience, then reviewing other situations that resemble yours can help. The interviewing of key people who know about your audience is also beneficial in helping you develop a profile of your audience's backgrounds. The more you understand your audience the more likely your message will "hit home." What follows in this chapter are some ideas to help in analyzing audiences and understanding how messages are received and acted upon, or—as is so often the case—ignored.

INTERNALS VERSUS EXTERNALS

Internals are people directly involved or identified with an organization. They can be expected to identify with the goals and missions of the organization producing the message. They also tend to have a vested interest with the outcome of any educational or informational message. Externals, on the other hand, have no vested interest in the organization and a message aimed at them must emphasize the "so what" question. Typical types of groupings within these separations are given in the table below.

Internals	Externals
Non-Governmental Organizations	
Full time employees	Community groups
Seasonal staff	Civic associations
Retirees	Consumers
Volunteers	Recreationists
Members	Business community
Board of directors	
Corporate donors	
Governmental Organizations	
Full time employees	Visitors
Seasonal staff	Neighboring Communities
Retirees	Special interest groups
Volunteers	Business community
Contractors	Non-elected leaders
Advisory boards	
Legislators	

POPULATION SEGMENTATION

Another way of viewing groups of people and how they will respond to new ideas and actions is through population segmentation. Like most audience analysis, these categories are just generalizations (audience profiles) on how a population may respond to a message. Actually knowing your audience is critical to success.

Adoptions of New Ideas

Adopter distributions in the American population follow a bell-shaped curve over time (Rogers and Shoemaker 1971; Rogers 1995). This segmentation of the American population describes how likely it is to adopt new ideas and technologies. A communicator will need to assess how "average" their target population is and in which categories they are more likely to be located. As a whole, the following categories give useful insights into a population.

- **Innovators** (2.5% of the population)—These venturesome few are the vanguard for new ideas and behaviors. They tend to initiate "new things" and take social risks in adoption of new ideas. They are most often wealthy and socially established with large ranging influence. They are also information seekers getting their information from primary sources, or may even be sources of new information. While they are "trend setters," they are rarely used as advice givers.

- **Early Adopters** (13.5%)—This segment is respected by others as being continually innovative and ready to try new things. They include both formal and informal leaders of organizations with social influence. They tend to be young and knowledgeable users of in-depth and specialized sources of information.

Both innovators and early adopters prefer hard evidence and factual accounts over unverified anecdotes. They also are the kind of people who seem compelled to have the newest "whiz-bang" gadget before anyone. They are truly trendsetters.

- **Early Majority** (34%)—This large segment is deliberate in their decision-making about new ideas and trends. They include many informal "quiet" leaders with influence within smaller social groups and small communities. They tend to be early to middle age with modest financial means. Though they are still opinion leaders, they are viewed as discriminating by others. They do not accept, or even test, every new idea they come into contact with. Their information sources are commonly family, friends, and mass media.

- **Late Majority** (34%)—A skeptical group, the late majority usually adopts only established ideas. They tend to be older with moderate education and "keep to themselves." They have conservative lifestyles and learn much of their information from acquaintances, friends, and family.

- **Laggards** (16%)—These people define the term "traditional." They adopt ideas reluctantly and only after they are unable to avoid change. They perceive great social risk in new things. They are often older people with the least education, lowest incomes and lowest social status. Friends and family of the same social station are their primary sources of information. Sometimes they ignore change agents and rebel if forced to try and adopt something with which they do not identify.

Support of Pro-Environmental Issues

Roper (1990) categorized the American public into five major groups based on their environmentally responsible behavior, awareness, and consumerism. Remarkably, political affiliation was not a factor in establishing these segments. Also, loci of control were predominately external for all (see Locus of Control later in this chapter).

- **True-Blue Greens** (11% of the population)—Like the Innovators, True-Blue Greens tend to be affluent and respected members of society. Interestingly, two-thirds of True-Blue Greens are women. They have the highest incomes, are more educated, and tend to be innovators within their communities. Their behaviors tend to be consistent with their strong concerns about the environment.

- **Greenback-Greens** (11%)—This group can be described as affluent with high incomes and substantial education, though below that of True-Blue Greens. They also demonstrate less environmentally responsible behavior. They are, however, large supporters of environmental organizations, through monetary donations. They will pay slightly more for environmentally friendly products, but do not usually become involved in environmental activism.

- **Sprouts** (26%)—As a swing group, they can be pro-environmental or anti-environmental on any particular issue. Their concerns wax and wane, depending on the saliency of the issue at hand. Overall, they are near-average in income and education. They show the characteristics of "typical middle America." While they have green tendencies, they show no enduring pro-environmental behavior. They are rarely willing to pay more for green products and almost never become activists.

- **Grousers** (24%)—Generally anti-environmental, lower income and less educated, individuals in this segment consistently rationalize their lack of environmental behavior by offering excuses. They also criticize the poor performance of others.

- **Basic Browns** (28%)—This group holds few opinions on environmental issues. Their behavior is not environmentally friendly. Their incomes are low. They tend to be poorly educated and are predominately male. Most work as unskilled labor. Of all five groups here, Basic Browns have the lowest levels of action and actually do not want to make any effort for the environment.

Adopting New Ideas

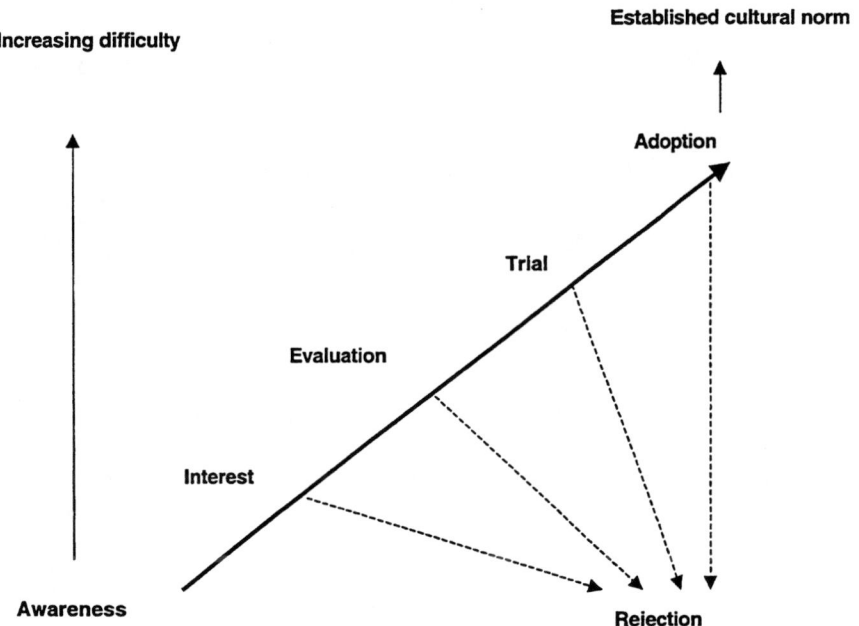

Figure 6   The Adoption Sequence

In the adopter categories above, the idea that populations can be segmented based on how they adopt ideas and technology was given. The process for populations is a little more complex, however, since different people move through the adoption sequence at different rates. Our culture is built on ideas that have been widely accepted. Thus, some ideas become the ideal and are called cultural norms. For example, we reject killing of others as wrong in our society. "Thou shalt not kill" has been upheld as a norm for millennia.

Ideas flow through societal discourse, either gaining acceptance or being rejected. The adoption sequence modeled here functions in a similar fashion as the knowledge filter. People develop acceptance to an idea that the sender has to give to the audience. It is not instant acceptance, and indeed, the adage "you can lead a horse to water but you can't make it drink" is most applicable here. It takes a lot of effort from the communicator to move an idea through an audience from awareness to acceptance, even when you have a receptive audience. As an environmental communicator, you seek to interject ideas into the discourse of your audience.

The adoption sequence theory (Lionberger 1960; Rogers 1995) posits that ideas go through a five-step process before the behavior of a large segment of the population can be affected. As an idea moves from one step to another through exposure to more people, the amount of effort by the communicator to maintain the process usually increases. Sources of information used also shift as people begin processing information for deciding whether to accept or reject an idea. Ideas may first come to awareness through mass media, usually via television and newspapers. But adoption is more likely to be driven by interpersonal contacts with respected and trusted friends. Factors affecting the adoption sequence are many and varied, and a wise environmental communicator will develop their plans based on sound evaluation and knowledge of the target audiences. Let's take a cursory look at the steps in the adoption sequence:

Awareness

This is the easiest step for communicators because it involves just exposing the audience to an idea. Most often, this is accomplished through mass media. The influence of friends and relatives is a secondary source of exposure and tends to be more expeditious, rather than planned. Awareness does not necessarily involve any changes in a person's attitudes and opinions. It also does not absolutely lead to retention of the idea or contemplation of its implications. Simply getting new information on a public's psychological radar screen is required, however, if further steps of the adoption sequence are to be attained.

Interest

Interest occurs when an idea resonates with an audience. To be interested, the audience members need to be shown the salience of the idea. For this reason, the process becomes a little more involved at this stage. Usually, this involves showing an audience how the new idea affects them in some specific way. Mass media are still the primary modes of communication at this stage. Still, a communicator is working on knowledge acquisition within the audience. Attitudes and opinions have not yet been affected.

Evaluation

This is a critical stage, for here the purpose of messages changes from knowledge transmission to persuasion. Mass media are now less effective as people begin to consciously consider the options presented within an idea. Advice from experts and other trusted opinion leaders is sought. Therefore, having credible sources for audience members to interact with becomes a crucial factor in maintaining an audience's progress up the sequence. People begin here to examine their own attitudes and opinions.

Trial

At this stage it is important to have positive reinforcement as individuals begin trying out the idea's upshots. If attitudes and opinions have been called into question, alternative behaviors will be tested. Experts may need to increase personal interaction with the primary audience, or develop messages for secondary audiences that can be influenced by the primary audience. Acceptance is made more likely if the individual has a positive connection with the idea or behavior. The idea or behavior may be rejected at this level if the benefits do not fit within an individual's belief structure (see below for more explanation of this concept).

Adoption

After successful trials with a newly-derived opinion, an idea, and its attendant behaviors, can be adopted. An idea can still be rejected, even at this stage. Since a communicator will have a lot invested at this stage, outcome or impact evaluation (see the next chapter) may reveal what was unacceptable to the audience. This assessment will allow future efforts to be modified to meet goals and objectives. If adoption takes place, it is important to continue positive reinforcement of the adopted idea, usually through the mass media. Continuation of messages prevents the audience from regressing to their old positions and galvanizes their new adoption. All in all, it takes some time for new behaviors and ideas to become established as cultural norms.

These levels can also be viewed as having unique functions associated with them. Awareness and interest are considered knowledge functions. Evaluation is considered a persuasion function while trial and adoption are decision functions (Solo & Rogers 1972). Knowing these functions helps the communicator focus a message for a specific purpose when viewed as part of the adoption steps.

Beliefs, Values, Attitudes, Worldviews and Opinions

In referring to how people think and act, the words "beliefs," "values," "attitudes," "worldviews" and "opinions" are commonly used. As one delves deeper into communicating about the environment, these terms take on special meanings. These meanings and their implications are important to state, so that later discussions about them can grow from a common understanding. Environmental communicators need to make many assumptions about their target audiences. Most of these assumptions involve the group's shared beliefs, values, attitudes, worldviews and opinions. The complex created by one's beliefs, values, and attitudes can be referred to as their "worldview." Those with similar worldviews are predisposed to act in similar ways. The best course a communicator can take in considering the reactions generated by their messages is to really know their audiences and to understand that beliefs, values and attitudes are culturally derived.

Beliefs

A belief is a basic, deeply-held ideal upon which people intellectually make sense of information. Beliefs structure what we think about other people, objects, and issues. Many times they can be expressed as simple declarative statements. Examples of common beliefs are: "Knowledge is good," "There is a God," and "Killing is wrong." Beliefs help us make judgments about information and lead to inferences about cause and effect (McNelly 1973). Beliefs offer a basis for mental pictures of what we hold to be real and acceptable (Rediker, Mitchell, Beard & Beach 1993). So, beliefs generate each person's perceptions of reality. These individual perceptions do not always agree (Petty & Cacioppo 1981). An individual's internal logic truly produces their view of reality. What is unquestionable fact to one may be pure fantasy to another.

Beliefs can be conceptualized as conglomerating together to form a mental framework through which all sensations received by an individual is filtered. These belief systems are large networks of beliefs that link different concepts with one another and give form and meaning to information (Rokeach 1968). They endow the holder's given state of affairs with sufficient validity and trustworthiness to warrant their reliance on their system as a guide for thought and action

(Harvey 1986). So, belief systems are self-sustaining and can be expected to change slowly, if at all. Belief systems are heavily reliant on an individual's experience and are, therefore, constructivist in nature. Belief systems also help us to make predications about the future by generating expectations about situations, actions, and other people (Paap 1989). In short, a belief structure acts as a template for making sense of the world and in deciding what is correct and what is not (Rumelhart 1980; Waern 1977).

Belief systems are also referred to as belief structures and schema. Regardless of the term used, belief systems are not random gatherings of beliefs that remain static within an individual's mindscape. Their structure is dictated by a connectedness that is logical, at least, to the holder of the system. We are all believers in our own way, you could say. This connectedness can be conceived of as constraints on what an individual will find true about reality. Horizontal constraint is where beliefs are related to each other in a chain-like sequence. For example, beliefs on chemicals, manufacturing, and big corporations tend to be linked together. Vertical constraint is where successive beliefs, often taking the form of a logical syllogism, are derived from each deeper belief (McGuire 1960; Bem 1970). Syllogisms are aspects of reasoning where one premise about something leads to a logical premise about something else and hence to a conclusion. For example, my friend Elaine reads the daily newspaper to keep her informed of world events. She is always well informed. Therefore, newspapers keep you informed.

All of these assertions about beliefs might lead you to conclude that changing opinions and affecting behavior is a foolhardy proposition. Communication that directly attacks one's belief system will almost always be ignored or interpreted as threatening. And, yet consider how much of the information you come into contact with daily is meant to change your mind on one topic or another— advertising, political rhetoric, religious pronouncements, etc.

Clearly, new information often does not squarely fit with our existing belief structures. Such information may be disregarded, "glossed over" and not mentally processed fully, or delegated to an area of uncertainty where conscious reasoning can still occur. Important in the mental processing of new information vis-à-vis belief structure is an individual's expectation about whether the incoming information needs to be fully considered or not. More on this process as we move from beliefs to their close relatives, values.

Value

A value is a personal benchmark of what is desirable. Values depend on beliefs and are more directly responsible for guiding everyday behavior. Like beliefs, values provide a framework for making decisions in specific situations (Kluckhohn et al. 1951). Whereas beliefs are truth-claims about what cannot be proven or fully attained, values offer standards by which individuals judge themselves and their realities. So, an enduring idea that a specific mode of conduct or end-state of existence is personally or socially preferable is a value (Rokeach 1973). From the belief "Knowledge is good" could come the value "More knowledge is better." In function, beliefs and values are closely aligned and are nearly inseparable when you think about them in designing communication campaigns. Again, attacking values will almost always get you nowhere with recipients of your messages.

Like beliefs, values are organized into systems. A value system allows its holder to envision preferred modes of conduct and desirable end-states of existence. In essence a value system is a learned organization of principles and rules to help us choose between alternatives and resolve conflicts (Rokeach 1973). Values are standards with which we judge claims and decide whether they are worth challenging, protesting, debating about, or ignoring. All experiences are examined using values. Predicting personal judgments by understanding an individual's values is not exact, but can be helpful in planning environmental communication.

Attitude

Information about specific objects and situations is framed and assessed through one's belief and value structures. This process of framing, and the inevitable drawing of inferences about an object or situation, is how attitudes are formed. Attitudes are focused and enduring mental organizations of information, based on several beliefs and values. They predispose one to respond in a certain way (Eagly & Chaiken 1993). Notice how attitudes are a product of beliefs and values. Sometimes understanding the underlying belief and value systems can help a communicator understand what factors may be hindering members of their audience from accepting a messages and making changes based on it. Notice also that attitudes are specific toward individual things within the reality of each individual. Attitudes vary subtly among members of groups.

Worldviews

When we look at different audiences, we need to be mindful of their sociocultural backgrounds and also many other aspects that define how they will look at the world. Worldviews are the composite of all our beliefs, values and attitudes that we have come to accept as our own. We all have a particular mindset that dictates how we filter information and make sense of what is happening. Our personal worldview also dictates what we expect to receive. When a message seems alien to this perspective, we will misinterpret the message, or just disregard it. This could be thought of as faulty encoding and decoding within the communication model. Worldviews are covered more in Chapter 15.

Opinion

When one expresses a belief, value or attitude, one has an opinion. Opinions are, therefore, the verbal evidence of an individual's worldview. Whereas beliefs, values, and attitudes are rarely stated openly and can be difficult to reveal, opinions must be made manifest. Opinions can be thought of as beliefs, values and attitudes put into words. Without some form of communicative representation, there can be no opinion (Rokeach 1968).

Situational Factors

Just as noise permeates and confounds communication systems, situational factors can short circuit, derail, and otherwise disrupt people's beliefs, values, attitudes, and opinions. Situational factors are all of those annoyances, unfortunate events, and disasters that arise continually to complicate everyone's lives: phone calls on your way out the door, flat tires, being broke, getting the flu, a death in the family, a flood, a stock market crash, a meteor strike, etc. Even peer pressure and social norms can be situational factors. From the standpoint of a communicator, situational factors get in the way of free flow of information. They are a form of noise.

Situational factors will interfere to make a person continue in a mode of behavior that is seen as "real-world" rather than the way they may want to behave. Sometimes a message may have to empower people to make changes they would like to make, rather than just informing them of a situation.

Locus of Control

Locus of control (LOC) is another psychological concept important in environmental communication. People perceive their ability to control the situations in which they are involved in different ways. A person who believes their actions can affect a situation's outcome has an internal locus of control. One who believes they are powerless in changing a situation's course has an external locus of control. Internal and external loci of control can be conceptualized as falling along a continuum. While in different situations the same individual may have different locus of control reactions, one's predisposition toward one end of this continuum or the other is moderately stable over time (Rotter 1954; Levenson 1972; Phares 1976; Lefcourt 1980).

INTERNALS (I control my situation)

- Believe their actions can make a difference
- Internalize control
- Account for only a small number of people
- Information seekers
- Take action in reference to their beliefs although such action is not legally mandated. Example: a person who practices recycling when it is not mandatory.

EXTERNALS (THEY control my situation)

- They either do not have the necessary information to take action or don't believe their actions will make a difference. May just require motivating or training in action skills to become internals.
- Need to be convinced of the importance of their role.
- Believe they cannot facilitate change, only "powerful others" can do so. Powerful others are perceived as having control and do not depend on any real group being active and noticeable.

When audience analysis unveils information about an audience's locus of control, it can be extremely helpful in planning a campaign. Internals on a particular issue may just need to be made aware of a problem. Externals need much more attention if they are to become involved. A small group of externals may fall close to a cut-off point on the continuum and only need some action skills and more information to generate activist tendencies. Many more externals will need to be motivated to act so that they gain ownership of the problems addressed. They may also need empowerment messages to become active. The group perceiving that "powerful others" exist may need extensive empowerment programs before they will even begin to consider listening or acting.

In summary, the communicator needs to understand that beliefs and values are extremely difficult to change. Indeed, most people are resistant to change.

Attitudes and opinions are expressions of these factors and make observation of the underlying cultural norms easier to fathom. This overall cultural mindset determines a worldview that the communicator needs to understand if they are to place a message to get a target audience to change something they have "always done." When the communicator is trying to get an audience to adopt new ideas, it is critical to know the barriers that exist within that audience to adopting something that will change their lives.

Understanding how the audience "thinks" is a crucial step in assessing how likely they are to respond to a specific message and how further messages need to be structured. As members of an audience advance up the adoption sequence, it is necessary to keep reinforcing the messages so that they do not reject them. Knowing why the audience is interested in the first place will help in avoiding the barriers that lead to rejection. Locus of control is important since people will respond if they feel they have the power to do something. Apathy is a major barrier that may be overcome by establishing ownership of a situation. Things that have a personal impact are more likely to encourage action. But it must be ensured that the audience feels it is empowered to act. The focus of the message will change completely, even about the same topic, depending on the psychological mindset of the audience. An audience that is highly empowered with an internal LOC and a profile of an early majority/greenback-green may need just information and encouragement to act. An audience comprised mainly of grousers or late majority with a LOC in 'powerful others' will need to be targeted for empowerment and motivation messages before any hope can be expected for adopting a new idea. Helping the audience understand how new ideas will impact their lives and then reinforcing the message demands a lot of work and a complete media campaign if success is expected. Knowing the audience is a major step in this process.

A Model of Citizen Participation

The model put forth concerning the learning of environmentally responsible behavior by Hungerford and Volk (1990) deals with three levels of activity that the communicator should consider when developing a message for a target audience. After considering the psychological factors already given, also consider each level of this model and each variable. Note how the focus of the message must also vary to address specific points of these variables. While some of these variables may be more important for any given audience, it would seem that all the "pieces" should be there before expectations of success are met. Like the adoption sequence at the beginning of this chapter, this model is another way of viewing the successive steps of communication planning that need to be considered if the overall goal is action.

Entry Level Goals + Ownership Goals + Empowerment Goals ⇒ Environmentally Responsible Behavior

Entry level variables include:

- awareness about the situation/issue
- sensitivity towards the situation/issue
- basic knowledge about the situation/issue
- have developed attitudes towards the situation/issue

Ownership level variables include:

- in-depth knowledge about the situation/issue
- a sense of personal investment (a stake)
- a personal commitment to resolve the situation
- knowledge of the positive and negative consequences of decisions about the situation/issue

Empowerment variables include:

- knowledge of action skills
- internal locus of control
- intention to act

The concepts of motivation, control, and individual needs are covered in the following sections.

It is paramount that message senders understand their audience. But audience analysis involves much more than mere characterization. Successful communicators must also understand their audience. This might be achieved first hand by talking to selected members of a target audience, or by gathering information on the actual behaviors of a target audience. If a communicator can reveal which beliefs might be threatened by the issue being addressed, then messages can be better designed and targeted. Opinions can be examined without producing outrage. Beliefs and values can be directly called into question or the campaign will fail. In the Hungerford and Volk model of Responsible Citizen Action (1990), the second step of "ownership" might be more essential to establish in an audience than the other steps, especially in light of the audience's situational factors.

Different people can conceptualize the same issue in radically different ways. People have attitudes that are narrowly focused and dependent on their immediate circumstances. Public opinion is often based on loose and shifting coalitions. Therefore, environmental attitudes tend to be issue specific and not necessarily continuous across all other issues. Mass belief systems usually have little depth or breadth to them, and are organized around more day-to-day and apolitical concerns unless galvanized by a specific current situation. During times of controversial situations that affect large groups of people, there are clear signals given within a community on what and whose interests are affected. Environmental protection consequently has often been a coalition of "pet peeves" rather than a unified movement (deHaven-Smith 1988). The selected attitude theories given here emphasize just how difficult it is to generalize from one population to another, or even within populations. Though it is not feasible to analyze each person in an audience, a professional communicator needs to build a campaign on an awareness of the psychological factors at play within their potential audiences. With probing and skilled audience analysis, it is possible to understand how to construct a successful message or why a message already in use may not be working. Audience analysis works hand-in-hand with evaluation (the topic of the next chapter). Both are crucial components to overall success of environmental communication.

Motivation

We have talked about motivation being an important factor in getting people to act. Motivation can be described as an individual's drive to get things done in his or her life. Motivation directs the formation of new opinions and transforms them into actions. As such, motivation is closely related to locus of control. People prefer to do what they are interested in, so messages that appeal to interest most often appeal to the audience's motivation.

Motivation can be typed as general, specific, intrinsic and extrinsic. General motivation is an enduring disposition to strive for new knowledge and skills (Brophy 1987). When tied to the message or sender, motivation becomes specific. For example, when someone you love asks you to do something, you have more motivation than if a stranger would ask you to do the same thing. Motivation that comes from within an individual is intrinsic. Other names for intrinsic motivation are curiosity, personal gratification, and growth potential.

When motivation is reinforced by external standards—test scores, salaries, degrees—it is extrinsic. Use of extrinsic motivators in communication campaigns has been shown to be risky. Meeting a standard or being compared to a larger group may motivate people to succeed, but fail to affect permanent behavioral change. The reinforcement is motivated by a fear of sanctioning and not necessarily a desire to succeed. Plus, giving extrinsic rewards for doing what someone is already interested in actually decreases their specific motivation (Calder & Staw 1975; Morgan 1984). When giving rewards, it is better to reward individual action rather than achievement of higher standards. People like to be compared with themselves, not faceless statistics derived from large groups. When using such reinforcers as motivators, care should be taken to ensure that the target audience does not become unmotivated because of misplaced efforts to promote action and learning.

The best motivation comes from within. Communication techniques that tap into people's intrinsic motivation can be highly effective. It is also difficult to produce messages that do this (Combs & Avila 1985). Humanistic approaches state there are no unmotivated people. The sender of a message must provide the relevant connections to the receiver. Answering the question "What's in it for me?" is key.

Motivational Needs Models

Our personal needs affect how much motivation we will invest in various tasks. Needs theory proponents stipulate that motivation occurs because people want to remedy deficiencies in their basic needs and to attain proficient access to higher needs or growth desires. The main implication of needs theory to the communicator is that people are unlikely to listen to a message that appeals to the aesthetic or higher thinking concerns before the basic needs of everyday living have been met. For example, asking people below the poverty line to "save the dolphins" by only buying higher priced dolphin-free tuna is not likely to have much affect. These people most probably cannot afford to support such a boycott, and may be too preoccupied with day-to-day living to worry about anything but surviving in their own lives. Understanding your audience's "needs" will help you develop appropriate messages. The three models given here emphasize the kinds of needs that need to be understood by the communicator.

Murray's (1938)/Atkinson's (1964) *Manifest Needs theory* proposes that multiple "learned" needs interact simultaneously and two factors, direction and intensity, drive the desire to satisfy the needs. Direction is the focus of the need e.g. a safe haven. Intensity is the strength of the desire to satisfy that need. For example, when only a mild threat exists there is little desire to hide, yet when a large threat exists there is a paramount desire to escape and hide. Some of the needs identified as manifest are achievement, competition, individuality, nurturance, order, power, and empowerment. What people need is often based on cultural upbringing and cultural expectations.

Maslow's *Hierarchy of Needs* (1943) is the most well-known needs theory and posits that humans have a hierarchical arrangement of needs from basic lower needs (deficiency needs) to higher existential needs (growth needs). These needs were expanded in 1970 to seven levels with the first four deficiency needs being Survival (shelter, food, water, and warmth), Safety (freedom from physical or psychological threat), Belonging (love and acceptance from others), and Self-esteem (recognition and approval, self-worth). The top three growth needs are Intellectual Achievement (knowing and understanding), Aesthetics (order, truth, and beauty), and Self-actualization (philosophical thinking, spirituality). The main premise of this theory is that an individual must first satisfy the lower need before they will advance to the next level to satisfy that need.

Alderfer's *Existence/Relatedness/Growth* (ERG) Needs theory (1972) posits two extra components of "frustration-regression" and "satisfaction-progression" that control the levels of need at which a person remains. Therefore a person acts on needs that are under frustration since these needs take precedence over ones that are already satisfied. When a need is satisfied, the person will want to progress further. If a need is frustrated, then a person will regress and prioritize that need until it is satisfied again. This does not imply that the other needs are disregarded, it just emphasizes that frustration needs will be answered before other needs.

All three of the above theories emphasize how individuals require their needs met before they will entertain change. If we consider the segmentation groups, then the early change agents, such as innovators, early adopters or true blue-greens, could be equated to the higher and already satisfied needs levels. This allows them to contemplate becoming involved in new situations. They are not preoccupied with the basic needs. Alternatively, the categories such as laggards or basic browns are concerned with everyday needs and are unlikely to find time or energy to devote to esoteric issues, which they also feel unempowered to affect anyway. Having a knowledge of the level of needs of your audience will also affect how they will respond to a message. If a specific unempowered audience is important to the outcome of a communication campaign, then the communicator will need to address the audience's needs before attempting to get them to begin adopting new behaviors.

How to Motivate Adults

With all of the motivational theory to work from, here are encapsulated directions for reaching adult audiences through your environmental communication campaign. (Reaching children requires different considerations, both practical and ethical.)

- Create or identify a current need in the audience and then respond to it.
- Get them to develop a sense of responsibility.

- Create and maintain interest within the audience.
- Structure messages to apply to real life.
- Give recognition, encouragement, and approval if they demonstrate action or interest. Praise and constructive criticism are best, rather than just praise.
- Foster wholesome competition and cooperation.
- Get excited yourself and then share your enthusiasm.
- Explain what is in it for them.
- See the value of internal motives.
- Intensify interpersonal relationships.
- Give them a choice. Let them be part of the process.

Ideas presented in this chapter emphasize the importance of honestly knowing your audience. Successful communication campaigns are built on such familiarity. It is important to know what level of knowledge and social action the receivers of your messages have. You should be able to explain their beliefs, values and attitudes, and the opinions they express. With such understanding you can motivate your audience. Environmentally responsible behavior can result. Know them well and you will get the response you expect.

REFERENCES AND FURTHER READING

Alderfer, C. P. (1972). *Existence, relatedness and growth.* The Free Press.

Atkinson, J. W. (1964). *An introduction to motivation.* Van Nostrand.

Bem, D. J. (1970). *Beliefs, attitudes and human affairs.* Brooks/Cole Publishing Co.

Brophy, J. (1987). Syntheses of research on strategies for motivating students to learn. *Educational Leadership, 45*(2), 40–48.

Calder, B., & Staw, B. (1975). Self-perception of intrinsic and extrinsic motivation. *Journal of Personality and Social Psychology, 31,* 599–605.

Combs, A., & Avila, D. (1985). *Helping relationships* (3rd ed.). Allyn & Bacon.

DeHaven-Smith, L. (1988). Environmental belief systems: Public opinion on land use regulation in Florida. *Environment and Behavior, 20*(2), 176–199.

Eagly, A. H., & Chaiken, S. (1993). *The psychology of attitudes.* Harcourt, Brace, & Jovanovich.

Harvey, O. J. (1986). Belief systems and attitudes toward the death penalty and other punishments. *Journal of Personality, 54*(4), 659–675.

Hungerford, H. R., & Volk, T. L. (1990). Changing learner behavior through environmental education. *The Journal of Environmental Education, 21*(3), 8–22.

Kluckhohn, C., et al. (1951). Values and value orientations in the theory of action. An exploration in definition and classification. In T. Parsons, & E. A. Shils (Eds.), *Towards a general theory of action* (pp. 388–433). Harper & Row.

Lefcourt, H. M. (1980). Locus of control and coping with life's events. In E. Staub (Ed.), *Personality: Basic aspects and current research* (p. 229). Prentice-Hall.

Lefcourt, H. M. (1982). *Locus of control—Current trends in theory and research* (2nd ed.). Lawrence Erlbaum Associates, p. 186.

Levenson, H. (1972). Locus of control and other cognitive correlates of involvement of anti-pollution activities. *Unpublished doctoral dissertation.* Claremont Graduate School, CA.

Lionberger, H. F. (1960). *Adoption of new ideas and practices.* Iowa State University.

Maslow, A. H. (1943). A theory of human motivation. *Psychological Review,* 50, 374–396.

Maslow, A. H. (1970). *Motivation and personality.* Harper and Row.

McGuire, W. J. (1960). A syllogistic analysis of cognitive relationships. In C. I. Hovland, & M. J. Rosenberg (Eds.), *Attitude organization and change.* Yale University Press.

McNelly, J. T. (1973). Mass media and information redistribution. *The Journal of Environmental Education,* 5(1), 31–35.

McQuail, D. (1997). *Audience analysis.* Sage Publications.

Morgan, M. (1984). Reward-induced decrements and increments in intrinsic motivation. *Review of Educational Research,* 54, 5–30.

Murray, H. A. (1938). *Explanation in personality.* Oxford University Press.

Paap, K. R. (1989). Applied cognitive psychology. In W. L. Gregory & W. J. Burroughs (Eds.), *Introduction to applied psychology.* Scott, Foresman and Co.

Petty, R. E., & Cacioppo, J. T. (1981). *Attitudes and persuasion: Classic and contemporary approaches.* William. C. Brown Co.

Phares, E. J. (1976). *Locus of control in personality.* General Learning Press.

Rediker, K. J., Mitchell, T. R., Beard, D. W., & Beach, L. R. (1993). The effects of strong belief structures on information-processing evaluations and choice. *Journal of Behavioral Decision Making,* 6(2), 113–132.

Rogers, E. M. and Shoemaker, F. F. (1971). *Communication of innovations.* The Free Press.

Rogers E. M. (1995). *Diffusion of innovators,* (4th Ed.). The Free Press.

Rokeach, M. (1968). *Beliefs, attitudes and values.* Jossey-Bass Inc.

Rokeach, M. (1973). *The nature of human values.* The Free Press.

Roper Organization, Inc. (1990). *The environment: Public attitudes and individual behavior.* The Roper Organization (New York).

Rotter, J. B. (1954). *Social learning and clinical psychology.* Prentice-Hall.

Rumelhart, D. E. (1980). Schemata: The building blocks of cognition. In R. J. Spiro, B. C. Bruce, & W. F. Brewer (Eds.), *Theoretical issues in reading comprehension* (pp. 33–35). Lawrence Erlbaum Associates.

Solo, R. A., & Rogers, E. M. (Eds.). (1972). *Inducing technological change for economic growth and development.* Michigan State University Press.

Waern, Y. (1977). Comprehension and belief structure. *Scandinavian Journal of Psychology,* 18, 266–274.

Yopp, J. J., & McAdams, K. C. (1998). *Reaching audiences: A guide to media writing*. Allyn & Bacon.

Youga, J. M. (1989). *Elements of audience analysis*. MacMillan College Division.

7

EVALUATING YOUR MESSAGE EFFECTS

Evaluation is the ongoing process of providing feedback about your messages' effects. Just like audience analysis and issue investigation, evaluation is indispensable to carrying out your communication plan well. Yet, evaluation is often the most neglected portion of a communication campaign. Many programs and plans often have no basis for determining how successful they were or for analyzing where problems may have occurred. Imagine attending a class where the only measure of success is when the teacher feels satisfied that they have taught well. Does that give a measure of how much the students have learned, relate good techniques to other teachers, or help the teacher improve instruction?

Many communication campaigns have been set up with just this sort of vagueness in order to justify themselves. Does knowing how many fact sheets were taken at a local nature preserve during a given weekend tell us how effective the fact sheets actually were in giving effective information to the users? Do the users actually read the fact sheets? How do we know? Answers to these kinds of questions are the essence of evaluation. An environmental communicator needs to genuinely want information that helps them understand problems and improve programs.

Think back to the communication model again. The element of feedback is equivalent to evaluation. A model depicted on a book page fails to show the dynamic aspect of feedback and the way evaluation can take place at most any stage as the system operates. In this chapter, we examine the need for feedback in the form of evaluation—why it is needed and how it can be done.

PURPOSES OF EVALUATION

Evaluation is a systematic process of judging program effects. It measures the worth and effectiveness of what is being done. A program in environmental communication could be an interpretative talk, a classroom activity, or a mass media campaign. Environmental communication messages seek to modify an audience's environmental behavior in some way. Whatever the particular modes employed by your program, there is embedded within it a need to check to see if it is working the way you supposed it would. A program that is not meeting expectations can be modified or terminated.

Evaluation has at least three other purposes worth considering. They are:

- To assist decision-makers who are responsible for deciding policy. People with policy-making duties need to know how effective a program is in order to wisely make decisions. Such decisions can be about the continuation of the program itself or about broader policies.

- To create accountability for resources used. When resources have been committed to a program, the communicator, as the program manager, needs to track and justify how those resources were used. Resources include budgets, personnel, equipment, energy and time. Resources, especially cash, are always limited. Decisions about allocation of resources need to be based on valid information generated from defensible evaluation.

- To serve a political function. This is probably the most surreptitious use of evaluation. A simple equation is: Evaluation = Information = Power = Politics. Politics is simply defined here as the art of exercising power in competition for resources and the setting of priorities. The idea of politics here covers both public and private domains of human society. There are elected officials and corporate agents with tremendous power. Simply stated, evaluation of communication campaigns can give a political player an upper hand in some situation of consequence. Good evaluation is timely and pertinent to this exercise. The people who provide such information have power to help or hinder the politicos. Providing the right information at the right time can make or break a program.

METHODS OF EVALUATING

Data for program evaluation are collected through several methods. The method selected will depend on the specific informational needs behind the evaluation. Cost also comes into play in deciding the way data will be gathered. When evaluative data are arranged and put into context, they become valuable information. Communication without evaluation is haphazard and, ultimately, wasteful of resources.

Here are the common methods of evaluation:

Surveys

Surveys use questionnaires or interviews to gather data. Whether they involve pencil and paper, computer, telephone or face-to-face dialogue, surveys are one of the most complex, misused and misunderstood methodologies of systematic information gathering. Yet surveys can yield highly significant data.

A problematic area is deciding whom to survey. Selection of the group to be surveyed, called sampling, raises many statistical questions. Randomness must be present in the selection process for your results to be generalizable to the larger audience. There are many ways to sample correctly, but there are even more ways to select a faulty sample and reach false conclusions. Still, no sample is perfect. An entire population would have to be polled to absolutely know.

Developing a reliable and valid survey takes much more than just jotting down a list of questions that seem relevant and addressing them to a group of audience members. Survey research requires careful structuring, testing, and administration. The aim is to ask questions that are "valid" (unable to be misinterpreted) and "reliable" (able to produce consistent responses). When pondering the use of a survey, consider the variety of types available.

- **Mail surveys**—A written questionnaire is sent via mail to a sample from a target audience. The biggest problem is to motivate those in the sample to fill out and return your questionnaire. Those that do respond cannot be assumed to be representative of the whole audience, without further checking into why they did not respond. While techniques exist for dealing with problems of non-response, it is essential to recognize the limitations it imposes on a mail survey.

  Another consideration of mail surveys is their self-administered nature. Those in your sample are asked to review their exposure to your messages and respond to questions about their reactions. This is a difficult feat to accomplish. People need to see the utility in helping you. Most times, you'll wish to provide them an incentive, some reward for filling out your questionnaire. Even with incentives, it may be difficult to get more than 50 percent of the individuals to respond. There is also a problem of determining if there is a difference between those people who respond to the questionnaire and those who do not. For example, it may be that those people who do not like or agree with the message will choose not to respond.

- **Telephone surveys**—Your sample is called using the telephone and, if they agree to participate, questions are asked directly to them. This method suffers most of the same drawbacks as the mail surveys. It also can be effective, if you can get past the barriers of answering machines, call waiting, pager numbers, etc., and actually get a real person on the line. As telemarketing has become more prevalent, the usefulness of the research-based telephone survey has declined. Telephone surveys can also be expensive, relative to other methods.

- **Face-to-face surveys**—Members of your audience are intercepted and questioned at a selected location. For example, this may occur at a museum gift shop, at the exit of a visitor center, or on the street in a mass media market where you have been running television advertising. The evaluator selects members of the audience to interview and then asks them to answers some questions. As with telephone surveys, a polite, pleasant and professional tone in which you respect the time and patience of the respondent will go a long way in motivating them to respond to you. This method can gather a lot of information in a short amount of time.

- **Personal interviews**—Similar to location surveying, except this technique uses fewer people in a sample and asks them more in-depth qualitative questions. Individual in-depth interviews often last between 30 minutes and one hour. The selected interviewees should still be representative of the target audience, though randomness will be more difficult to achieve. In-depth interviews are costly and time consuming, yet can yield superb insight on reactions to messages, issues, and policies.

- **On-site testing**—Small groups of the target audience (usually no more than 25 people) are brought together to review the various communication messages being developed within a plan. The messages need be set up in situations similar to the final delivery. In trying to decide on the effectiveness of different kinds of messages, it may be necessary to avoid telling the audience about each message they will be viewing. This will avoid some members of the audience giving responses about what they perceive you want to hear. For example, a public service announcement may be embedded in a half-hour television or radio program. In order to get useful information about the effectiveness of the announcement, it would be useful to not tell the test group about this message, and instead concentrate on the program in which it is embedded, and see if they picked up on your embedded message.

- **Comment forms**—Simple written questionnaires, usually formatted onto a post card, can be left at strategic locations to be picked up by interested members of your audience. Many restaurants will leave one at each table to garner simple feedback from customers about quality of the food and staff. Similarly, comment forms left near the exit of a visitor center can give quick information on how visitors liked or disliked the center's exhibits and activities.

Participant Observation

Sometimes, much can be learned by watching what a person does, rather than having them tell you about themselves. The evaluator takes on the role of a visitor, viewer, reader, or other participant. This allows data to be gathered from the perspective of the audience. The evaluator discretely observes audience behavior and also notes their own reactions. This is a particularly useful technique when one seeks to compare beliefs, values, and attitudes, with observable behavior.

An example of this might be in a visitor center where the observer records how people look at the exhibits, how long they stay at each station, and other relevant information to gauge just how effective the exhibits seemed to be in holding the audience's attention. Later interviews could be employed to give results on why some exhibits held attention while others were just dismissed. Another scenario might be where a community has begun a curbside recycling campaign. The observer might drive through the target neighborhoods on trash collection day and gauge how much trash versus recyclables was left by the residents. Later interviews could yield insights into why some people recycled a lot and others very little.

Interviews

Studies of the adoption sequence, as discussed in Chapter 6, as it relates to the American populace, have shown the existence of innovators who consistently try new ideas before most others. Asking members of this population segment about their reactions to your messages can be a wise move, if you are attempting to shift public opinion. Here are three ways to consider discussions with opinion leaders:

- **Opinion leader interviews**—Many government officials and business leaders can provide relevant information about your issue, target audiences, and messages. Often, they can be interviewed to elucidate needed information.

If many key designates are desired, then one of the consensus methods may be appropriate (see below).

- **Gatekeeper review**—Gatekeepers are the key people in organizations who decide what information is released to the organization's members or constituents. Prime examples are the editor of a newspaper or the communication/human resources director in a large company. Gatekeeper review involves soliciting gatekeepers for their perceptions of your messages. If gatekeepers do not respond well to a message, the target audience will never get a chance to receive it. For this reason, it is wise to include gatekeeper review as a method of evaluation.

- **Discussion with key informants**—Key informants are people in a community or organization who have unique insight into your audience or situation. Such people can be interviewed to gain specific insights into a program or issue. Examples of the sorts of people in this group may be the owner of a local store, a street vendor, a local restaurant owner, a local barber or hairdresser, and regulars at a local coffee shop or bar. No avenue for gathering information should be ignored. Key informants are not always the "leaders," yet in their own right are respected by the rest of the group for their wisdom and understanding.

Group Consensus

When a small group that reflects your intended audience is brought together, either actually or virtually, and is guided toward consensus by a skillful facilitator, group consensus technique is being practiced. Group consensus technique is useful in needs analysis or problem definition situations. Here the group can provide insights into what needs to be evaluated or to help define the problem more succinctly such that evaluation is more focused on gaining information to resolve the problem. Other uses of group consensus are for ongoing reviews of a communication project so that modifications can be made, or simply to do a final evaluation of a project using a representative group.

Here are four group consensus techniques to add to the environmental communication toolkit:

- **Brainstorming/Focus groups**—The brainstorming technique works on the Gestalt principle that the sum of ideas from individuals interacting in a group is greater than the sum of the ideas generated by individuals alone. This works best with around 10 individuals who are representative of the target audience. While there is research suggesting brainstorming is not the most effective method for generating ideas, such a session can be a fruitful, low-cost way of generating ideas about message effectiveness (Diehl & Stroebe 1991). In most cases, if a group has experience interacting with each other already, then brainstorming tends to be an enjoyable experience that stimulates a greater variety of ideas than working as individuals.

    Brainstorming can be facilitated by following these steps:

    1. Have all participants express their ideas, as you write them down for all to see. It doesn't matter how silly or crazy the ideas are, for the more ideas that are expressed, the better the process works and all ideas and information will be evaluated later.

2. Review the resulting list stressing that all ideas belong to the group. Encourage additional ideas stimulated by this review.

3. Discuss each idea, without reference to its originator. By focusing on relevancy to the issue at hand, the list can be prioritized. Take the best options and integrate them into your communication planning.

When 10-30 individuals are convened as a group and a guided through a discussion by a facilitator, you have a focus group. The facilitator allows the group to talk freely about the message, channel, and format. Focus group interviews have potential to provide insights into the target audience's perception of the issue and message effectiveness. Care must be taken, however, not to interpret findings of a focus group interview in a quantitative nature. For example, if five out of ten of the interviewees feel a policy is unacceptable, it does not mean that fifty percent of the target audience will agree. The process used to deal with a focus group can be brainstorming if the group is comfortable with each other. Otherwise, the following techniques of nominal group process, or Delphi technique may have to be used.

- **Nominal group process**—This is possibly the most efficient small group method for generating useful ideas. Individuals work alone initially and then come together as a group. This allows people to develop their own understanding of the issue at their own pace without distractions. Then, as a group, they are facilitated through these steps:

  1. Individuals write individual responses.

  2. Each person states one idea, which is recorded by the facilitator for all to see. Do not allow judgmental statements during this procedure.

  3. Participants are encouraged to ask clarifying questions and to suggest similar items which can be combined.

  4. A key word in each remaining item is underlined. Each item should have a different key word.

  5. Each participant selects the most important item from the list and writes that item's key word on an index card.

  6. Each participant ranks the resulting key words in order of importance to themselves.

  7. Rank orders are averaged so a democratically generated list results. This list should be checked by the group, with the help of the facilitator, to assure that it reflects group consensus.

  Note that no one openly discusses the value of any of the ideas in nominal group process. In this way, antagonistic responses are reduced to a minimum.

- **Delphi technique**—In some cases, it is not feasible to convene a group in a single location. If the people from whom you want to gather information are distant from each other or cannot find time to meet, then the Delphi offers an excellent alternative technique. In other cases, you may wish to not have people come together because of polarized opinions or the need to remain anonymous. Again, Delphi is a good choice. This technique avoids the problems of intimidation, co-optation, and peer pressure.

In the Delphi method of group consensus building, the evaluator drafts a series of questions about a topic, issue or program. This list is forwarded to each member of the group for their review and return. The evaluator then summarizes these responses and returns them back to the Delphi group for further feedback. This send-review-return-compile cycle continues until the evaluator is satisfied that some consensus has been reached.

Secondary Analysis/Case Study

Many times you can search for and find another program addressing similar issues to your own. Though it will be in another location, the details and travails encountered during planning and implementation are likely to correlate with your own. By reviewing what techniques were used and what findings where derived in this situation you may save a lot of time and money. Such a technique is called secondary analysis or case study.

TYPES OF EVALUATION

Evaluation takes place throughout the course of an environmental communication campaign. The timing of your assessments can be used to classify four types of evaluation. This distinction is taken from the Bennett model of evaluation (Bennett 1976). Although this evaluation model is widely used by many in the environmental communication field, it is only one of many available models. Here are the types:

Formative

Evaluation that takes place during a campaign's design and early in its implementation is labeled formative. Such evaluation is meant to gain information about the audience, head off potential problems, and help confirm the correct channels have been selected. There are several types of formative evaluation, including:

- **Context** evaluation, which assesses the information preferences and needs of the target audience.
- **Input** evaluation, which assesses which channels and formats that are most appropriate to reach a target audience.
- **Program** evaluation, which addresses the effectiveness of the actual transmission of the message before it is sent out to the main audience. This is also referred to as **pretesting**.
- **Needs assessment** is where "needs" are identified and prioritized for further analysis.
- **Needs analysis** is where identified needs are analyzed and solution strategies developed.

Process

While a program is in full swing, process evaluation can be used to learn about effects on the fly. That is not to say the resulting information is light and fluffy, however. Process evaluation allows rich and full descriptions to emerge as all aspects of a program are scrutinized. One benefit of process evaluation is problems uncovered have the chance of being remedied before a campaign is completed. The major drawback of process evaluation is its cost. Continually assessment of a program can become the most expensive part of your budget.

Outcome

As soon as a program wraps up, outcome evaluation can be conducted. It gives quick answers to questions about program effectiveness. How did it go? Did it make a difference? Outcome evaluation produces everything from simple head counts of the people reached by your messages to in-depth studies of the shifts in audience beliefs and behaviors. Usually this more sophisticated information is required to tell if your objectives and goals were met. A good outcome evaluation usually implies that a pretest was done during formative evaluation. This allows comparisons to be made and shows what changes have occurred within the audience.

Impact

Long-term effects of an environmental communication campaign are revealed by impact evaluation. This type of evaluation needs to take place after the completion of the program, often after several weeks. It measures the durability of the effects. Whereas outcome evaluation might show a large shift in audience perception, impact evaluation will determine if the change was sustained. If support is shown to have fallen substantially, a new program of simple reinforcement of the original message might be in order.

FACTORS INFLUENCING EVALUATIONS

One of the first questions that must be asked as you begin evaluating your program is: "What kind of data is needed?" Do you need "quick and dirty" results to help make a simple decision? Or will a "thick rich description" of what is happening be all that can adequately address your concerns? Coupled with these informational needs are the cost of the evaluation and the required timing of the results. These factors will influence the form and shape of your programmatic evaluation:

Cost

You cannot do what you cannot afford. Evaluation can become expensive. If the information you wish you had is financially out of reach, what cost-effective alternatives will provide similar information?

Expertise

Evaluation is a discipline unto itself. There are professional evaluators that specialize in complex projects of assessment. Do you have the required expertise to properly conduct the evaluation necessary for your campaign? If not, where do you get this know-how?

Reluctance

Before enacting any evaluation within your program, be sure that all those who will use your findings really want to know how effective the program is. Do not ask questions about awareness, influence, and behavior change if you do not want to know real answers. Elaborate, well-crafted, and expensive campaigns sometimes fail miserably. If finding out that such a fate has befallen your project is more risky than not having answers to questions of effectiveness, do not bother evaluating. Recall one of the purposes of evaluation is "to serve a political function." Know beforehand how poor performance would jeopardize your standing with the powers that be.

Sample Make-Up

Talking to the right people is important in most any endeavor. Evaluations are no different. There are statistical concerns as well as social ones in selecting your sample. Try to avoid upsetting members of your audience by leaving them "out of the loop." Involving many people throughout all aspects of the program is a good way to make them accept ownership of the program.

Utility

Results from evaluation efforts must be used or they are just a waste of time, effort and money. While an elaborate evaluation may be desirable and the funds to perform it are available, it is critical to know from the outset just how the results will be used.

Timeliness

Evaluations should be planned with deadlines attached. Getting information in time to act on it, if necessary, is the real value behind evaluation.

Autonomy

Because evaluation can just as readily show that resources are being wasted on a program that is accomplishing nothing as it can show success stories, evaluators must be careful to have autonomy. Independence from influence is paramount to any research effort, and evaluation is a form of research. As an evaluator you have to be equally prepared to share positive and negative results. You generally will have little problem when your results affirm that a program is functioning as planned. But, in evaluation, especially one conducted by those within an organization, you may feel pressure to find the "right" results. Be cautious about such intimidation. Keep your integrity and report your findings honestly, regardless of the influences on you to do otherwise.

Evaluation can tell you many, many things about your program: what works, what does not, why your campaign succeeded, why it failed, who heard what you had to say, who did not, etc. This feedback can make you a better communicator if you pay attention to the lessons your findings offer. Evaluation, though often neglected by communicators, makes your program more valuable by heightening its accountability and giving you solid evidence of what it has accomplished.

EVALUATION PLAN

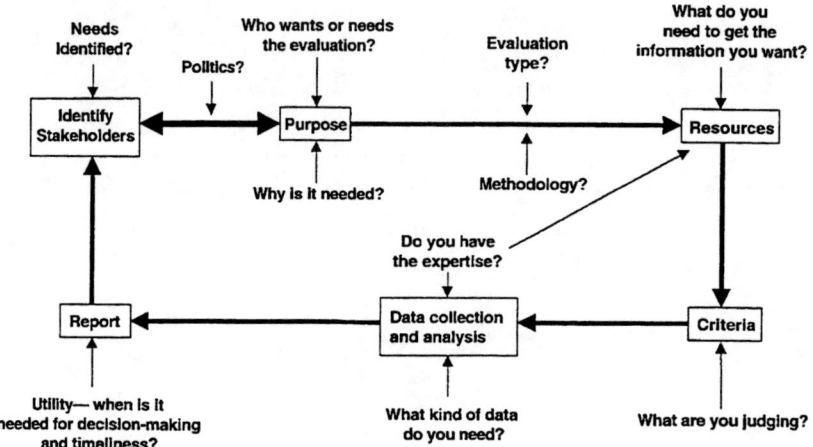

Figure 7   Evaluation Planning Schematic

In our schematic of the evaluation process, the top left corner of the diagram shows stakeholders and purpose. Stakeholders are the people with a vested interest in the evaluation. It is here that primary decisions are made about the evaluative process. This drives the kind of information required to satisfy the need. Sometimes a needs assessment and needs analysis must be the first steps to identify what is required to be evaluated. The next related step is to decide the purpose of the evaluation. This involves deciding who wants or needs the information and why it is needed. The concept of politics needs to be considered here. Once the process is decided, it is necessary to ponder on the mechanics. What kind of evaluation is needed? Does it require formative and/or summative evaluation? What methods will you use to collect the information? This will be completely dependent on the resources at hand. Examples of this are the budget available, and the expertise of the personnel in being able to collect and analyze the information. The decision on what to evaluate then necessitates setting the criteria for what information you need to collect. Setting objectives for the evaluation will guide the process. Data collection can be difficult and they can be even harder to analyze. If the type of data you require needs complex statistical analysis, will an outside expert be required? This brings us back to the review of resources. Once data are collected, the drafting of the report is critical. Is it going to be finished in time for the stakeholders and decision makers to use? If the evaluation is too complex it may take time to analyze and draft up a report. If an early decision was needed, then the report is redundant and the evaluation a wasted effort.

By moving through the different aspects of this schematic the communicator will be able to identify key components that could hinder the evaluation process. It is always pertinent to remember the factors that influence evaluations as you design the evaluative process. Each evaluation is unique. One size does not fit all. Once all the process questions have been satisfied, it is likely a useful evaluation will result.

REFERENCES AND FURTHER READING

Altschuld, J. W., & Witkin, B. R. (1999). *From needs assessment to action: Transforming needs into action strategies.* Altamira Press.

Bennett, C. (1976). *Analyzing impacts of extension programs.* Sage Publications, Inc.

Diehl, M., & Stroebe, W. (1991). Productivity loss in idea generating groups: Tracking down the blocking effect. *Journal of Personality and Social Psychology, 61,* 392–403.

Dillman, D. A. (1978). *Mail and telephone surveys.* John Wiley & Sons.

Fink, A., Bourque, L. B., Fielder, E. P., Frey, J. H., Oishi, S. M., & Litwin, M. S. (1995). *The complete survey kit,* vols. 1–9. Sage Publications.

Isaac, S., & Michael, W. (1995). *Handbook in research and evaluation: A collection of principles, methods, and strategies useful in the planning, design, and evaluation of studies in education and the behavioral sciences.* Edits Publishers.

Patton, M. Q. (1996). *Utilization-focused evaluation: The new century text.* Sage Publishers.

Scriven, M. (1991). *Evaluation thesaurus.* Sage Publications.

Weiss, C. H. (1997). *Evaluation.* Prentice Hall.

Wholey, J. S., Hatry, H. P., & Newcomer, K. E. (Eds.). (1994). *Handbook of practical program evaluation.* Jossey-Bass Publishers.

Worthen, B. R., Sanders, J. R., & Fitzpatrick, J. L. (1996). *Program evaluation: Alternative approaches and practical guidelines.* Addison Wesley Publishing Co.

CHARACTERIZING THE MASS MEDIA

Communicators today have a bewildering array of media choices for transmitting messages. Even the tried-and-true term "newspaper" might mean a city daily, a neighborhood or rural weekly, a national daily such as the *Wall Street Journal* or *USA Today*, an alternative weekly, a school newspaper, or any other regularly printed, newsprint-based publication. Advances in computer technology have created opportunities for communication far surpassing those of just a decade ago. Too often, a mode of communication is chosen without analyzing whether that particular medium is best for achieving the communication's goal. If your planning process has been thorough, then choosing among media choices will be based on your goal and objectives, target audiences, message characteristics, and budget. All of these are made explicit in a well-done communication plan. Do your job well in formulating the plan, and media selection will be smooth. Fail to write a sound plan, and you will be confused and dazed by the myriad of choices out there.

In this chapter, we will analyze the communication media, characterizing them not by type: newspapers, magazine, newsletters, billboards, radio, television, interpretive talks, etc. Using such a typology now strains its usefulness. Distinctions among types have blurred and computers connected to the Internet have expanded the modes available dramatically. Is an on-line magazine more a magazine or a digital news service? Are advertising panels in shopping carts more like billboards or display ads in print media? Compiling an inclusive list of types of media is no longer possible.

Instead, we will look at communication media based on qualities that cut across these outdated classifications. These characteristics include a medium's purpose, audience focus, delivery mechanism, timeliness, and cost. Each of these characteristics can be evaluated separately by a communicator when selecting channels for their environmental messages.

PURPOSE

It may sound cynical, but the main goal of most media is to make money. Most newspapers and magazines, for example, make their profits by selling advertising. Analyze your local newspaper by calculating the amount of space taken up by advertisements in comparison to the amount containing news. You might be amazed at the amount of space set aside for ads. A healthy newspaper is filled mostly with advertising and not news. In order to sell this much advertising, news media need to be able to demonstrate that their product is consumed by audiences that the advertisers wish to reach. Just like environmental communicators, advertisers have well-researched concepts of the groups of people with which they want to communicate. Audiences have specific characteristics and each news media outlet offers an unequaled group. For the largest of the mass media—the television networks—viewership size and demographics are demonstrated during "sweeps" weeks and through Nielsen ratings. Newspapers and magazines closely track their circulation and readership demographics. Most news organizations work to sell their audience to advertisers at least as hard as they work to sell their product to consumers.

Even those media distributed by non-profit organizations, or those that are free to the public, usually must make enough money to cover production costs. Public radio and public television must raise funds to cover those costs not subsidized by their corporate contributors and the government. Free publications usually come from corporations or non-profit groups that fund their communications through indirect means, such as advertising or sponsorship. Internet provider services usually have a price for subscription, are provided by a company or school, or send lots of advertising to their subscribers.

The financial goal inherent in most media is important to consider when analyzing whether they are appropriate for a specific communication program. For example, newspapers are interested in those stories that will sell more newspapers, and thus will be reluctant to cover a story that will be of interest to only a few readers, regardless of how important the communicator believes the story is. Likewise, television networks are unlikely to run a public service announcement for a non-profit group during a time when there is a high demand for advertising spots, such as during Monday Night Football or a popular situation comedy.

While the main goal of most media is to make money, they generally have one of three other purposes: information, persuasion, or entertainment. These purposes are highly interrelated, and often difficult to distinguish from a cursory analysis. But, knowing this additional purpose can help determine whether the communication mode is appropriate for your message.

Providing Information

Many media provide information in the form of news. Not all information is news, however. Chapter 16 covers the news process in detail, but it is important here to recognize that news is information that is important, interesting, and timely. Environmental communicators have access to news media in several ways. First, news releases are routinely used to announce activities of an organization, such as new hires, new initiatives, and public events. Box 1 describes the mechanics of writing news releases. Communicators can also help to create news in order to gain news coverage. Consider organizing a demonstration, putting on a sporting event as a fundraiser, sponsoring an expert or celebrity to speak, or inviting reporters to join your organization's workers as they accomplish an important task. A third means of making your work into news is to provide information that helps the news outlet do its job better. Offer yourself or an expert from your organization for an interview on a pertinent topic. Send your area reporters a packet of background information describing what you do. Write letters to key reporters in your community explaining what you do that might be newsworthy. For each of these approaches to be effective, the communicator must have a thorough understanding of the news process, and be able to frame information in a way that is interesting to the reporter, editor, and final reader.

Other media are used to provide information. For example, scientific journals are highly specified publications where scientists report the findings of their research. Many organizations publish various documents intended to provide information on specific issues or events. One example, the fact sheet, is a self-contained document that provides background information in an easily understandable format. Fact sheets can be found all over the place, and versions are now prepared specifically for the Internet.

Persuasion

Media meant to persuade range from speeches of those running for political office, to World Wide Web sites posted by advocacy groups, to handwritten opinion letters. All persuasive communication tries to get audiences to undertake a specific action.

Opinion letters are perhaps the most common example of persuasion. Generally, they are a personal appeal sent to an elected official, an agency official, or a business in order to achieve some desired effect. Letter writers can be highly effective in influencing legislation and rulemaking, especially when combined with other writers in a well-focused campaign. Other media whose purpose is persuasion include advocacy magazines, billboard and placard advertisements, public service announcements, and newspaper editorials.

Entertainment

A large number of media are concerned primarily with entertaining their audiences. While these media may appear useless for communicating environmental messages, creative communicators can use entertainment to achieve their goals. For example, the Earth Communications Office works behind the scenes in Hollywood to infuse environmental messages into television series episodes and movies. Their public service announcements have been spliced onto feature films and distributed throughout the world.

Environmental interpretation, as discussed in Chapter 2, often contains a large dose of entertainment in order to attract non-formal audiences. In addition, movies, television programs, cartoon books, and music can provide important messages in support of environmental communication campaigns. For lower prices, items such as T-shirts, bumper stickers, and novelty items can be creatively employed to achieve communication plan goals. Entertainment is a powerful tool when employed creatively.

Communicators have differential access to different media. For example, environmental communicators have limited access to feature newspaper articles in that a reporter and editor have control over what gets written, where the article is placed, and when it appears. A newspaper advertisement, on the other hand, is able to be controlled more by the communicator, since the space is purchased and the advertisement created either by or for the communicator.

AUDIENCE FOCUS

When choosing among communication modes, communicators must carefully match their target audiences with the audiences of the candidate media. Some media, such as newspapers, have wide, general interest audiences defined more by geography than any other factor. Alternately, many insider newsletters have small, diffuse, narrowly-focused audiences. No medium reaches everyone with a particular set of demographics.

An example of highly specialized media is scientific journals. They can be important sources for environmental communicators seeking new knowledge working its way through the knowledge filter (see Chapter 3). In scientific journals, articles are peer-reviewed, meaning that scientists in the field review all articles submitted to a journal, make suggestions for improvement, and accept only those articles that are deemed to exhibit high quality research, meaningful findings, and of interest to the journal's readership. Audiences for scientific journals are almost entirely made up of scientists working in the discipline covered by the journal. Some highly respected journals, such as the *New England Journal of Medicine*, *Nature*, and *Science* are read by wider, albeit technically trained audiences, including practitioners and reporters from mass media outlets.

Access to scientific journals is through direct submission by author-researchers. A peer review process generally involves one to four reviewers who evaluate the article without knowing the identity of the author or authors. All articles published in scientific journals require an abstract, and some journals publish only abstracts. An abstract is a short summary of an article, document, talk, or presentation. Box 4 shows the format of a standard journal article and the mechanics of writing a good abstract.

An example of media with wide audience focus is news magazines, such as *Time* and *Newsweek*. Millions of people, from all walks of life, interests, and demographics regularly read these magazines. While they may seem out of reach for many environmental communication campaigns, large conservation organizations often get their issues covered by these magazines. Recently, media mogul Ted Turner was on the cover of *Time*. The photograph showed him wearing a tie with The Nature Conservancy logo. That's good publicity from an article not focused on conservation.

It is important to be careful in analyzing audiences for media. The World Wide Web may seem like a widely focused medium with the potential to target millions of computer users. But, because of the way users search for web sites that interest them, even sites that appear to have thousands of visitors may have little holding power, and thus actually have only a small, narrow audience.

Knowing the audience focus of media is critical to implementing a communication plan. Careful audience targeting of messages is achieved through careful choice of communication mode.

DELIVERY CHANNEL

The delivery channel is the actual transmission form used. Newspapers are delivered on paper, while radio broadcasts use airwaves. In the past, media were often categorized as either print media or electronic media. Today, however, this distinction is blurred. Magazines and journals are now found on the Internet, as are live news broadcasts. Compact discs include written text as well as sounds and motion pictures. Technology has greatly expanded the range of delivery mechanisms.

In analyzing delivery mechanisms for a given message, the communicator first must consider appropriateness. For environmental messages, paper-based media may be inappropriate since many paper technologies are not sustainable and are wasteful. Some media may be considered unsightly. One may not want to run a beautification campaign with billboards.

At the same time, the communicator must consider other purposes for, and the long-term impacts of, various media. Paper-based media can be saved and used at later dates to document the communication. Electronic communications are often fugitive. People rarely record radio broadcasts, or even television broadcasts, and World Wide Web sites are changed often and are rarely permanent. Magazine reprints have an apparent legitimacy compared with screen prints of a now-changed web site.

In the field of environmental interpretation, the delivery mechanism can be as important as the message itself. Creative use of signage, slide programs, costumed characters, and actual objects can ensure a message gets to the intended audience.

The delivery channels of various media should be considered during communication planning. In combination with audience focus and media purpose, the channel used can have an important impact on the success of a communication project.

TIMELINESS

Media have various time-oriented characteristics that are important in communication planning. These are lead time, frequency, and longevity. Each of these impacts the delivery of a message to the target audience.

Lead time is the amount of time required to get the message to the audience. For example, a group that holds a public demonstration or a press conference may well be able to get their message on the evening news programs if television reporters cover the event. This same event might also be in the next morning's newspapers. While the demonstration or news conference itself may require many weeks of planning and organizing, once held the time to reach the audience is measured in hours. At the other end of the spectrum, some magazines, particularly those whose purpose is entertainment or persuasion, may have lead times of many months. Articles for magazines are often assigned to writers a year or more in advance. Communicators should consider these lead times as part of the implementation planning for their campaigns.

Frequency refers to the number of opportunities afforded by a particular medium for re-transmitting a message to a specific audience. This is not necessarily the same as how often the communication mode is distributed. A newspaper may be printed daily, but the communicator may be unable to get informational articles into the newspaper daily. On the other hand, advertisements can be purchased in daily editions of the newspaper for reinforcement of the message. Magazines are often kept for years after their delivery. Communication planners must determine to what extent message repetition will improve its comprehension by the target audience. Also, one must decide whether such repetition must come from the same communication mode.

Longevity is the amount of time the message will be available to the audience. Published materials generally have a lengthy longevity, especially if, like newspapers, magazines, and journals, they are retained in libraries and databases. Future accessibility can be an advantage when the message is one the communicator wants available in perpetuity. On the other hand, if a mistake is made, or if a past message is in conflict with a current one, this longevity can be a major drawback. Traditional electronic media usually have a short longevity, unless someone in the audience records the radio or television broadcast in which the message was contained. E-mail and World Wide Web may have short or long longevity, depending on whether the message is maintained on the Internet, or is lost (and becomes what is called "fugitive"). Communicators must match their strategy to media with appropriate longevity for effectiveness.

Cost

As discussed in Chapter 4, communication plans are implemented with a financial budget. This economic reality constrains the media choices available for delivering a message. Costs can include direct purchase of the communication mode, such as the purchase of television advertisement time, as well as the cost of creation, such as the cost of hiring a public relations agency to create the advertisement. An additional cost is staff time to manage the communication. For example, a news release may be free to publish, but writing, editing, and distributing it takes staff time that has a real economic cost to an agency.

As a rule, the cost of media increases with the size of the audience reached. So, newspapers with larger circulations have higher advertisement prices, likewise radio and television stations with larger audiences have higher advertisement prices. For direct communication modes, such as direct mailings, larger audiences have larger costs, although economies of scale can lower cost per piece mailed. Communication planners can, by dividing the target audience size by the total communication budget, calculate a cost per audience unit figure. Usually, media buyers compare this as a "cost per thousand," that is the cost per 1,000 members of the audience. Their abbreviation for this measure is "CPM." It can be used to assess the relative merit of different media within the total communication project.

Media costs also increase with elegance. For example, the cost of a simple fact sheet can increase dramatically as the number of colors in its design increases. Also, typesetting, the use of photographs and other graphics, paper weight and finish, and printing process all impact the cost of the fact sheet. The important point for communicators is to determine how much elegance is required for the message to be received by the target audience. Additional colors in a fact sheet may or may not add to the impact of the message. For each communication mode chosen, the communication planner must determine how much to invest in the medium to achieve, not the best, but the appropriate delivery of the message.

CONCLUSION

When choosing a communication mode, the environmental communicator tries to find the most efficient way to transfer a message to the target audience. The effectiveness of the message will be higher when the purpose of the medium is consistent with the message content and the objectives of communicator. Additionally, the communicator can accurately target the audience by analyzing the audience focus of potential media, and choosing the best fit. Delivery channel, timeliness, and cost constrain the possible choices for message transfer by excluding audiences or exceeding the means of the communicator. The choice of media is an iterative process within the planning process, as described in Chapter 5. The next chapter will highlight some of the more common modes of media that can be used.

REFERENCES AND FURTHER READING

Beamish, R. (1995). *Getting the word out in the fight to save the Earth.* John Hopkins University Press.

Calvert, P. *The communicator's handbook: Tools, techniques and technology.* (2000). Maupin House.

Miller, J. D. (1986). Reaching the attentive and interested publics for science. In Friedman, S. M., Dunwoody, S., & Rogers, C. L. (Eds.), *Scientists and journalists: Reporting the science as news*, pp. 55–69. The Free Press.

Parker, L. J. (1997). *Environmental communication: Messages, media & methods.* Kendall/Hunt Publishing Co.

9

HIGHLIGHTING USEFUL MEDIA

In the last chapter we addressed the importance of characterizing the media for best presenting a message to a target audience. While a list of media is endless, there are a few that seem to be used more than others. The list below shows some of the more common media. In this chapter we give suggestions on how to use them successfully.

1. News releases
2. Writing letters
3. E-mail
4. Abstracts
5. World Wide Web (WWW)
6. Public service announcements (PSAs)
7. Information sheets
8. Science writing
9. Direct mail
10. Unusual media

1. NEWS RELEASES

The purpose of a news release is to inform news media of routine events that reflect functions of organizations. These might be the election of a new president, or an annual event pertinent to a community. Non-routine events may also be of interest if they are timely and of broad interest to the news media readership. Occasionally, a minor crisis within an organization that affects a community may be best reported via a news release by the organization's staff writers who can ensure that the story is correctly given.

Mechanics of News Release Writing

A news release needs to be written on official organization letterhead. Also given in the heading should be a statement of when the information is pertinent for printing such as "Immediate release" or "Release on (date)." A contact person with address and phone number is essential.

The release should open with a descriptive headline that concisely presents the major emphasis of the article. This opening paragraph should cover the who, what, where, why, and how of the event headline. The lead, presented in active voice, gets the reader's attention.

The body of the release, written in inverted pyramid style (see Chapter 16), provides facts without editorializing. You cannot assume the reader has background information such as location, addresses, function of an organization, knowledge of abbreviations used within the organization, etc. Proofread your work to ensure relevance to the audience, timeliness, and correct grammar. These are critical to getting it accepted by the editor.

Common Mistakes in News Release Writing

Editors throw out the majority of news releases they receive, because the releases:

- Aren't tied to a local situation
- Are not newsworthy
- Contain too much advertising fluff
- Are long and cumbersome
- Arrived too late and are no longer timely
- Contain redundant information
- Are poorly written and/or presented
- Contain a source which cannot be confirmed or information which is suspected to be incorrect.

2.   LETTER WRITING

One of the most persuasive and powerful forms of communication that an individual can do is based on personal communication—writing a letter! The advent of modern electronic forms of communication seems to have made the humble written letter seem all but obsolete, but in reality the opposite seems to be happening. Electronic personal communications media (e.g. e-mail) is becoming the more commonplace mode to communicate speedily, but it has a cold impersonal feel to it. Even with the array of emoticons (symbols included to express emotions), the electronic messages are designed for fast, succinct transfers of information. And that is often how it is treated by most busy people. To really make an impression that time and effort went into sending a message or opinion, the clearly hand-written or typed letter is still a number one choice. Whether it be on a typewriter or a computer word processing program, the personal letter will still have an effect.

Not all letters need be letters of complaint or addressing a problem/issue; some can be affirmation letters of support. Whatever your opinion, unless you write and tell a specific someone about it, you will never be heard and the potential recipient will never be able to appreciate it or act upon it. However, before you write, do your homework. Know what specific action or situation you are asking the letter recipient to act upon. Or, if a product is the focus, define it specifically. Know what you intend to write about before you actually do the writing.

Why They Are Important?

A personal letter of opinion nearly always has an effect on the recipient. Simply because most people do not actually write their opinions! Most people will readily voice an opinion, but fewer will actually take the time to write it down and send it off. While estimates vary widely, it can be assumed that each letter that is written and sent also corresponds to at least 100 people's opinions that were never sent.

Letters to Legislators

An opinion letter does much more than just express your opinion, it acts as a barometer of public opinion in general for the recipient. There are different levels of legislative structure such as village, township, town, city, borough, county, state, federal. The state and federal are often subdivided further into administrative, judicial, and specific agencies. All levels are subject to various different kinds of bureaucracies. Despite the skepticism toward government officials, it must be assumed that they really do want to do the right thing as determined by their constituents. To this end, they need to hear from constituents about their concerns.

Be sure to write to the lowest level of legislation that can deal with your concern before moving higher.

Levels of Legislation

There are many levels of legislation and it is important to recognize at what level you want your letter to be read. You need to know the jurisdiction of the authority that you focus your opinion on. Who has the authority or capability to act on your opinion? It is also important to act on a chain of command. Begin with the lowest (most appropriate) level of authority able to deal with your letter and if you get no action or satisfaction at that level, write to higher authorities, and be sure to detail the lack of action from the lower parts of the chain.

Letters to Editors

Letters to the editor are another major option for expressing your opinion. The target audience this time, however, is the readership of a particular printed medium. Newspaper and major news magazine editorial letter sections are pulses of public opinion. Hence, legislators and policy-makers will also get a sense of the prevailing winds of public opinion from these sources. This channel may also prompt more people like yourself to act on an issue or appreciate something of value you have expressed.

Letters to Manufacturers/Businesses

These types of letters let manufacturers and businesses know where they are, or are not, succeeding. Their success depends entirely on good consumer reaction to their products or services. Any poor reactions can result in much negative response being spread by dissatisfied consumers, while positive reactions can help promote positive responses. It behooves an organization to maintain good relations with its consumers and to mitigate and remedy problems before they become damaging. This is also the essence of public relations. Your opinion letters of affirmation or complaint are a primary source of feedback for organizations that offer a product.

Business Writing, e.g., Memos, Letters

Business writing, like any other, demands succinctness. Try to place yourself in the position of the message recipient. Would you want to read this communication if you were busy?

There are five questions to ask of yourself before you send anything out:

1. Is this communication necessary?
2. What is the purpose of this communication?
3. Who is my reader? (If there is no audience, why are you sending it out?)
4. What are the relevant points to cover?
5. How should my reader respond to this document?

Then pick one of the five ways to sequence content for a business correspondence: chronologically; by priority; by problem, cause, and solution; using comparisons and contrasts; with advantages versus disadvantages. Whatever sequence you use, keep it short and to the point. That will ensure it gets read.

3.   E-MAIL

This mode of communication is one of the fastest growing forms of communication today. It allows you to communicate with people worldwide with the push of a button. While e-mail between friends may be longer, it must be emphasized that any business or non-personal e-mail really needs to be to the point. People who receive extensive numbers of e-mail messages a day soon stop reading them, or just scan the first few lines at best! In a sense, writing skills on e-mail are even more necessary if you wish the recipient to read your transmission. While emoticons (such as ☺ or ☹) can help to add some personality to your short messages, it must be remembered that email is a cold medium, and as such is subject to misinterpretation. People usually tend to send-off email messages without much editin.g. Sometimes it is wise to let your message sit for a while and then re-read it before sending. Much grief can be omitted by this simple practice. It has been said of more than one person in a senior position: "this person should have a delay and re-read key on all their out going e-mail, so that they stop sending out messages that keep putting a foot in their mouth." While it is a great way to reach a lot of people quickly, please don't intrude on people just because it is easy to do so. Be polite, reasonable, and succinct!

4.   ABSTRACTS

In today's fast-paced and media-rich society, most people are overwhelmed with too much information. It is expected that any lengthy document or scientific article have a summary attached that is complete enough to give the reader an accurate overview of the material. This synopsis, usually called an abstract, should also entice the reader to want to read further into the material. Obviously when time is a constraint, as it is for many professionals, the abstract may be the only part of the document that actually gets read. Therefore, it behooves the writer to carefully write this to ensure that the material is to be considered at all.

What Is an Abstract?

It is an essential condensation of the major points. It is non-evaluative, providing a succinct overview of the contents. The purpose is to provide the reader with enough information to determine whether the abstracted document is relevant enough to read in its entirety. Essentially, the abstract is a time-saving device. It is an essential condensation of the major points that clearly overviews the topic and conclusions. It does not reflect the opinion of anyone other than the author of the original communication.

Mechanics of Writing Abstracts

A good abstract is accurate, self-contained, concise and specific, non-evaluative, coherent and readable. It should open with a lead sentence that sets the stage and explains the topic of the original article. The body should contain no more than a couple of hundred words, preferably on one page, in which all of the major concepts of the main material are brought to the reader's attention. It should end with a one-sentence conclusion that pulls the abstract together and helps the reader evaluate the original document.

5. THE WORLD WIDE WEB

The World Wide Web (WWW) is fast becoming popular for "connected" people in finding information and presenting information. While it is true that the WWW is an excellent source of information and also a way for you to present information, it is fraught with several problems.

- The design of a web page is crucial. The opening page must contain all the relevant information without the need to scroll. Buttons for "links" to other parts of the website must be visible as well, with enticing captions. The page design is open to creativity, but should not distract from the message. The main drawback is that thousands of new pages are added every day, so unless people know your specific web address, they may not find you.

- First, there is no consistent organization on the WWW as there is, for example, with a library. Finding the information you want can be time consuming and hit-or-miss. Getting to a useful site can be purely overwhelming, with the amount of information you do find. While there are hundreds of search engines to help you find information, nearly all use unique search strategies to locate key terms within a web document. Just becoming familiar with web browsers and search engines can help a lot, and by using several different search engines, you can learn to locate more obscure pages. Using different search strategies of your own to thin down the search selection field will also help in attaining a manageable search list to review.

- When a webpage author develops a web site and then publishes it to the WWW, it must be understood that unless people have that page's unique web site address, they may not be able to find it through a search! Many search engine companies have "netbots" (automatic search programs) that continually scan web documents for key terms. Key terms contained in a coded line under "meta" headings are more easily found by these netbots. However, this can take weeks to months, if the document is ever found at all. Most web site authors usually have to submit their site to the search engines if they wish to be found on web searches. However, unless you have selective key terms, you may find that your page is found among many millions of other pages! Many on-line submission companies offer a service that both submits your website to hundreds of search engines and also implants techniques to make it appear at the front of any search list when your key terms are entered.

- **Validating information** is essential. There is presently no gatekeeper or peer reviewing of information that is placed on the WWW. An old adapted term from consumerism is apt here: "information user beware." Just because the information is on the web doesn't give it any validity. Things to think about when reviewing a web site are:

Enormous amounts of information can be found on the WWW. Some good—some not so good. Some is research data. Some is just unsubstantiated opinion. Some is to provide information. Some is for persuasion. So what can we do to judge the good content from the not so good?

Evaluating a site means using your judgment of the veracity of the content. There can be good information on "bad" sites and bad information on "good" sites. Some can even be blatantly false. Information on the Web should be viewed with the same skepticism as information from "regular" information sources e.g. fact sheets, newspapers, etc.

Questions to Ask of Web Site Information (EETAP, 1999)

- **Authority**—who wrote the information? Are they credible? Where was the information derived from?
- **Audience**—Who is the audience? What is the focus of the web site?
- **Context**—Why is this information being presented to this target audience? Is there any obvious bias? Is the information broad or deep? What links are there to other pages?
- **Accuracy**—Is the information "sound?" How do you know? Are there any sources given that can be verified?
- **Currency**—Is the information recent? Is it updated as appropriate?

6.   PUBLIC SERVICE ANNOUNCEMENTS

Radio is a great medium to augment an ongoing communication program. One easy mode of using radio is with a public service announcement (PSA). These are short announcements aired throughout the day using odd time slots during regular programming. Non-profit groups and agencies are the users since these slots are free. The Federal Communications Commission (FCC) requires radio stations to give air time in return for a license. But the stations can choose what they air and when.

Radio stations are keen to air messages that are noteworthy, promote a cause or service they consider important, or inform the listening audience of something special. It behooves you to get to know the station managers to find common interests between the station and your organization. This then allows you to tailor your messages so that your PSAs appeal to the station and the listening audience. You might send in various lengths of your script so that it might be played in those odd time slots that unexpectedly crop up. Suggested lengths would be 15, 30, and 60 second long scripts.

In submitting your PSA you may also send in a taped version along with the script, but usually the station will use one of its own announcers who will just read your script. The announcer is often the disk jockey in which the convenient time slot occurs. To ensure that your message is announced clearly and correctly, the following list will help the announcer know exactly what you want to be said and how. Grammar rules for PSAs are relaxed. You are writing to be heard and must quickly create an image in the listeners mind. You can even use fragmented sentences as is often found in conversational speech.

- Plan on writing a conversationally paced script. A guideline is to have about 25 words per 10 seconds of air time.

- When you need a pause use three dots, or more if a longer pause is required.

- This is preferred over semicolons and colons. Use a double hyphen to indicate an abrupt change.

- Use an identifying phrase or explanatory title if beginning the message with an unknown name.

- Difficult names or words must be followed by a phonetic spelling in parentheses. For example, with the African name Thato Nadayitwayeko (Ta-Toe Na-Day-It-Waa-Yea-Koe) this makes pronunciation easier. Do not use parentheses otherwise. If an unusual name or word can be pronounced different ways then use phonetic spelling or underline the syllables where the accent falls. For example, the town of Waukesha would be pronounced correctly as (Wau-Ke-Shaw) or Wau<u>kesha</u> and not Wau-kes-ha.

- Do not use abbreviations unless they are everyday ones. If you do, then use hyphens between each letter to let the announcer know each letter is to be pronounced separately. For example, U-S-D-A prevents the announcer saying "usdah." And, not everyone knows what USDA stands for. Do not abbreviate the names of places or address identifiers such as street, avenue, or boulevard. Neither abbreviate calendar terms such as days of the week and month, or the titles of officials. However, you may use Dr., Mr., Mrs., and Ms. with a name.

- Use "st," "nd," "rd," and "th" after days of the month, such as February 1st, April 2nd, July 3rd, or October 10th.

- Phone numbers can be written as you would normally write them with hyphens, e.g. 999-555-1212. All other numbers should be numerical to 999, but then alphabetically written for the zeros. For example, 89 and 247 are fine, then write 9 thousand, 21 million, 67 billion, 237 trillion, or even 25.6 trillion. Similarly when using fraction such as ¾ write out three-quarters.

- Avoid sibilants—the "s" words—that cause hisses and whooshes on the microphone. Also avoid using alliterations—tongue twisters—not everyone can say them easily on a first reading.

- If you have an address or a phone number in your message, have it repeated two or three times at the end so listeners will more likely remember it, or use an easy to remember mnemonic.

- When finished, have a colleague speak the script out aloud in the timed length you have specified. This will help catch the errors and other problems an announcer may come across.

7.   INFORMATION SHEETS

Information sheets convey succinct amounts of information and are not meant to be comprehensive texts on a topic. They come in a vast array of sizes and formats. These can range from a simple one-sided fact sheet to a multi-page brochure. Whatever the format used, some simple features will make the format attractive, the information appealing to read, and importantly, ensure that the target audience picks up the material and reads it. Regardless of the type of information sheet used, the reader should be left with some form of action to pursue. This might be to phone or e-mail a contact source for more information, or at least be given a series of references or further reading to find out more about the topic. Inclusion of WWW addresses will also lead the reader to more sources of information. Not giving any outlets of this kind will leave the reader frustrated if they want more information. Make it easy for them to follow an interest you may have generated in them. Consider not only the information, but how the target audience will obtain and use this information.

Considerations for any information sheet are:

- **Format**—Deciding whether to use a simple fact sheet, a folded brochure or a multi-page booklet, as just three examples, depends on the need of the information given and the audience for which it is targeted. Assess how the audience will use this information and under what conditions. For example, an informative fact sheet to be mailed to homeowners, needs to be suitable for mailing. It also needs to be concise so it will be read, and light enough so postage costs are minimized. A three-panel pamphlet would be suitable to carry in a back pocket for someone on a short self-guided nature trail. A multi-page brochure would be useful to someone spending a few days exploring the flora and fauna of a wilderness refuge. However, backpackers intent on keeping their packs light will forgo a bulky brochure regardless of how informative it may be.

- **Color**—Decide if the sheet will be a single color type on a plain background, or many colors. Printing costs can be prohibitive for color documents. If the budget is limited, then monochrome printing may be the best option. However, black print on white background can be appealing if layout considerations are observed.

- **Layout**—The title should be bold and stand out clearly from the rest of the text. It should also convey at a glance exactly what the information sheet is about. Breaking the page into columns or panels makes the information easier to read and gives the impression of "sound bites" rather than diatribes of the topic. The font chosen needs to be easy on the eyes, large enough to read, yet allow the inclusion of suitable amounts of information. Graphics should be clear and easy-to-understand. Each graphic should be closely associated with the corresponding text in the layout. Alignment of text, margins, borders, shading, graphics, section separators, and judicious use of "white space" all support an appealing layout.

8. SCIENCE WRITING

Scientific journals are not approachable to those without specialized training. Even scientists in one discipline can find a journal from another to be bewildering, the language foreign. Non-scientists can hardly be expected to ever want to look at a scientific journal. So, taking scientific findings and translating them for broader audiences is a highly marketable skill and one vital if society is to have broad scientific literacy.

A bit less than half of the American public pays attention to scientific news (Miller 1986). Many of these interested persons actively seek news of scientific discovery. Writing stories for them takes special skills. Honing an ability to construct stories from science is the bailiwick of science journalists. Gleaning story ideas from primary sources, the scientific journals, science writers then interview scientists in a manner that respects the tentativeness that is hallmark in the culture of science. Scientists as sources for news walk a tough line between public service and ostracism among their professional peers. Generally, they are not trusting of journalists. A reverse is also true: journalists often find scientists difficult and unrealistic in their expectations of the news media's role in society. They both can develop better relations by focusing on their common ground: the diffusion of knowledge.

The best science-focused news results in wider participation in policy-making, as informed citizens express their opinions as to how scientific findings can be incorporated into the society's workings. Intermediaries can play catalytic roles in the generation and dissemination of important science news. Examples of such intermediaries include university public relations personnel, government public affairs workers, non-governmental activists, and many other professionals who have interest in spreading findings from research.

9. DIRECT MAIL

A frequent method for targeting specific audiences is direct mail. It can be as simple as a posted fact sheet, or a collection of various media. Thus, it is a technique rather than a specific medium. In this age of mass mailings, it is easy to think of this format as junk mail, but it is efficient and cost effective. It targets the audiences of interest with a message pertinent to their needs. It also lets the audience follow up easily with contributions to a cause, requests for information about an issue, inquiries about products, etc. In essence this is a form of social marketing (see Chapter 19). Effective direct mail demands credibility, which includes a viable cause to support, products that appeal to the audience and match their needs, and of cause current and valid mailing lists that match the message and audience. It is more than just information, it strives to get a reaction from the audience. Some examples of direct mail are:

- Non-profit groups who solicit contributions to maintain their advocacy and lobbying efforts.

- Allows non-profit groups to alert an audience about arising situations and develop support. These types of messages can be easily personalized to make the audience feel appreciated.

- Special events can be easily promoted. For example, an Earth Day rally, or community clean-up days.

- Updating a target audience about a developing situation. Marketing ecological safe products via mail order to a geographically broad audience.

10. UNUSUAL MEDIA FOR ENVIRONMENTAL COMMUNICATION

Most anything you can think of can be used to deliver a message: public hearings, conferences, action days, field trip, rallies, open houses, art contest, eco-tours, street fair, celebrity photo-ops, church programs, youth groups, benefit concerts, teacher workshops, demonstration, trail walks, puppet shows, T-shirts, balloons, bumper stickers, door hangers, utility bill inserts, parades, comic books, coloring books, pens, pencils, erasers, place mats, posters, exhibits, baseball hats, coffee mugs, toys, flags, songs, poems, museum programs, story books, billboards, murals, photographs, curriculum materials, software, electronic bulletins, slide shows, video games, cartoons, field guides, information kits, speakers bureaus, calendars, music tapes and CD-ROMs. The list really is endless. All can become suitable channels for messages with the help of a competent environmental communicator.

It is key to remember that most of these alternative formats should be support media, not primary message conveyors. Use alternate means for simple messages in support of more complex communications via more conventional channels. In conclusion, it is essential that the communicator select and design the media that will best address the purpose of the message and reach the target audience. The cost, focus, and timeliness will all dictate what can be accomplished. The type of channel chosen and the medium used will vary with each audience, so plan to be flexible.

REFERENCES AND FURTHER READING

Beamish, R. (1995). *Getting the word out in the fight to save the Earth.* John Hopkins University Press.

EETAP, 1999—*Evaluating the Content of Web Sites.* (1999). Environmental Education and Training Partnership (EETAP) Library, OSU Extension, 700 Ackerman Road, Suite 235, Columbus, OH 43202-1578.

Miller, J. D. (1986). Reaching the attentive and interested publics for science. In Friedman, S.M., Dunwoody, S. & Rogers, C. L. (Eds.), *Scientists and Journalists: Reporting the science as news,* pp. 55–69. The Free Press.

Parker, L. J. (1997). *Environmental communication: Messages, media & methods.* Kendall/Hunt Publishing Co.

The communicator's handbook: Tools, techniques and technology. (1996). Maupin House.

Zehr, J., Gross, M., & Zimmerman, R. (1991). *Creating environmental publications: A guide to writing and designing for interpreters and environmental educators.* University of Wisconsin—Stevens Point Foundation Press, Inc.

SECTION 3:
SKILLS BUILDING AND
PRACTICAL APPLICATIONS

10

GROUPING TOGETHER WELL

A "group" forms when two or more people interact in a way that allows each person to influence and be influenced by each other member of that group (Shaw 1981). Interaction differentiates a group from students sitting in a hall listening to a lecture, movie-goers viewing a film, and passengers being carried by an elevator. These other assemblages lack the dynamics required to be considered a collection of people influencing each other, to be a real group.

The form of interaction within a group usually determines the types of motivation, group agenda(s), and interactive roles that exist therein. In this chapter, the broadest classification we'll use distinguishes formal from informal groups.

- **Formal**—Organized to do jobs with specific end goals. Formal groups are usually established within permanent organizations with enduring types of work and hierarchical command structures. Such groups often handle tasks and assignments within the overall structure of their parent organization.
- **Informal**—Formed by members who share a common goal or interest. Informal groups usually are more ephemeral than formal groups, often focused on resolving specific problems.

Environmental issues offer many interesting case studies of group formation and transformation of informal groups into organizations containing many formal groups. Many non-profit groups were created to deal with specific issues of concern to the founding members. Later they expanded to include many related interests.

WHY DO GROUPS EXIST?

Groups form for reasons of shared need. A group's members wish to get something from the group that they are unable or unwilling to get for themselves. Attraction to groups falls into two categories.

Primary motivation operates on an interpersonal level. People with primary motives join a group because they are attracted to qualities of other people within the group, have a vested interest in an outcome, or identify with the group's shared beliefs, attitudes, and opinions. Members generally seek opportunities for more personal interaction with other members. Such social identification with a group serves to enhance the ego-function of the individual. There is a direct task orientation when there is primary motivation.

Secondary motivation occurs through less strong feeling. This type of involvement serves as a means to an end, where the group meets a pragmatic, finite purpose. Joining a group for secondary motives allows one to network with desirable others, creating career-building or social circle opportunities. The prestige of membership—and not what the group could accomplish—is what is important. Colloquially, such membership can be referred to as "padding the resume."

COMMUNITY GROUPS AND THEIR SPECIAL ASPECTS

Environmental issues often create community groups. Citizens with primary motivations in seeing their environment protected organize for action. These groups tend to be short-lived and highly focused on the resolution of a problem. Their purpose is often singular and their tasks few. But, members can help themselves to be more successful if they pay attention to several qualities. A group's ability to overcome the negative qualities makes problem resolution more likely. Such group dynamics also shed light on the possibility of a more formal organization and permanent existence. While the communicator may be a central figure in the organization of the group, it is imperative that the group become self-sufficient. The following factors emphasize special points that a communicator needs to be aware of when helping organize community groups.

Hegemony

Hegemony is the act of imposing a set of values on a situation or group. Though this concept is usually applied in the domain of international relations, it can also be fruitfully applied to group dynamics. Whenever people interact, each agent brings their own particular mindset and belief system to the situation. Forceful imposition of one's position is a form of dominance and implies that the others' positions are of no consequence. Such coercion may only serve to alienate and close lines of communication. It is important, as a communicator, to realize that your values are yours. Do not assume others automatically share your values, your beliefs, your attitudes, and your positions. Even if you are convinced that your views and values are the "right" ones, it is up the community group to discover it for themselves and therefore, incorporate those values and beliefs into their own cognitive framework.

Empowerment

A hegemonic view of group power says that empowerment, the ability to get the job done, comes from a higher authority. Some stronger group gives power to a lesser group. This view is most often counterproductive, because it is debasing and degrading. In contrast, a thoughtful communicator will assist a group in the discovery of its particular avenues of actions. Such a remedy will be more satisfying and more resistant to failure. Metaphorically, show them the map and let them find a path, with your guidance if necessary. Give them the skills and help them find a way, through their own experience, to find a solution.

Revelation

Discovery is a process by which individuals in a group come to understand concepts or events from a new perspective. It is the role of the good communicator to help people construct realizations for themselves. In the field of environmental interpretation, Tilden (1957) defines this essence of discovery as "revelation." Understanding is ideally revealed to a person or group such that they see connections for themselves without the need for outsider explanations. While understanding the sense of what is being communicated is certainly important, it is the process of discovery that imparts deeper meaning to the message.

Education as Intervention

All too often, education becomes a process of telling people what the educator thinks they should know. This ultimately can lead to indoctrination of dominant beliefs that gives a distorted view of needs from a dominant perspective. It is critical to the success of any educational effort to first be aware of what the target audience perceives its needs are before imposing education on them. Concepts from social marketing (see Chapter 19) and risk communication (see Chapter 18) are important here. Educational interventions must square with the community's existing perspectives (Heimlich & Norland 1994).

Leadership and Dependence

Groups tend to form a dependence on a leader and so need to learn self-reliance and interdependency. Leadership should, when possible, be grown from within a group. If a leader must come from outside, then this person may have to work hard at overcoming resentment and a feeling of intrusion. Cohesiveness of a group tends to be better maintained through an inside leader. But, if an outsider is able to become an accepted member of the group, their leaving can result in more powerful feelings of abandonment and potential for dissociation. Therefore, an outsider who gains leadership of a group must try not to become the central decision-maker of the group. Instead, the outsider should lead as a guide and not as a ruler, especially if the group must continue to act long-term.

Openness

The degree to which interested persons are allowed to be involved in group process can be crucial to the success of a group effort. Any organizations that use public interaction must do so with the full intention of listening to what the audiences have to say! Audiences that are given the carrot on a stick of being involved and then find out theirs is just token involvement can quickly become hostile and non-cooperative. It is critical that you let the audience know at the outset what its role is in the interaction. Is it merely information-giving, in which they will have no power in the decision-making? Or will they be a full partner with the ability to say "No." The audience's role must be clearly delineated for them.

Team Building Techniques

One of the givens for a professional communicator in today's world is working in "teams." This can be either within an organization, a loosely-formed coalition, a task force, or an informal community. We've all heard horror stories about having to work in such groups, but the experience need not be so negative. In this section, we offer some ideas on how to build a strong team that takes into account the characteristics of individuals and actually uses them to forge a stronger consensus-building group. After all, isn't that why we all come together in groups?

Groups and organizations are part of how society has structured itself. People long ago made the decision to affiliate in a myriad of forms. Consider how many groups you belong to both within your job environment and your personal life. Consider the pros and cons of group membership:

- **Positives:**   Ideas and information generate better products
  Comfort and stress reduction
  Spreading of the workload
  Tasks can be achieved more quickly
  Different experiences and ways of thinking contribute to new ideas and solutions
- **Negatives:**   Hard to organize
  Hegemonic tendencies
  Uneven distribution of work and resentment
  Often no compromise situations develop

Groups are not simple entities to function within. But, the issues and task they deal with are not simple either. Gestalt theory states that you get more from the whole group than you could from the sum of the individuals. While in most cases this is true, some team members may work better if left to themselves to work on the problem before inviting them to discuss it within the whole team. Knowing the preferences of the whole team will direct which method is best.

Group Climate

Groups function within an emotional and psychological environment. The dynamics of the group are heavily affected by this group climate. A climate conducive to profitable dynamics and to reaching satisfying outcomes can be helped by clear rules of participation, permissiveness, and cohesiveness. Who is allowed as members, what is allowed and prohibited within group interactions, and expectations of how the group will work together are best stated up front. Groups also produce a better climate for themselves to function in if they keep close tabs on member motivations, have agreement on their means of reaching agreement or consensus, and are vigilant in managing conflict.

As noted, motivation can be central to a member's affiliation or quite tangential. Leaders and communicators working with groups should develop some sense of the motivational makeup of the group. Having this sense permits efforts to be more focused, by hanging on the needs and desires of group members. Likewise, having a known decision-making process in place makes groups function more efficiently. Will formal rules be followed? Will votes be taken by secret ballot? Will decisions be made only after consensus is achieved? A member who knows the answers to these questions will be better able to take part.

Perhaps the most dangerous component of group dynamics is conflict. Conflict is inherent in any issue. It should be expected between and within all groups. But, conflict can be used positively to enhance group development and foster progress toward solutions. Not dealing openly with conflict inhibits group function, generates resentment and hostility, and may destroy relationships.

Building Relationships

Instead of succumbing to inevitable conflict, groups that flourish find their strengths and weaknesses and then make these obvious to all members (see activity below). Groups are wise to make time early in their existences to identify shared characteristics and differences. This is particularly important because in most cases the group has come together because they feel they have a shared reality consensus on either an issue or a task orientation. What comes out of such an exercise is the revelation that although they might share the same larger conceptual ideal, they have very different reasons for being there, and even different agendas for working on that task or issue.

Differences occur because of values, beliefs and opinions which are often masked behind symptoms like too much focus on process and petty bickering. When we are faced with insufficient information to make a decision, we fill in the gaps from our own perceptual background and experiences. In essence, our own view of reality dictates how we perceive the world and we assume erroneously that others see the world in the same way. We need to accept that we hold different agendas. Such acceptance fosters cohesiveness.

Here are some suggestions on group rules which have worked many times before:

- The best number of members for small group interaction is five to seven. Groups with more than seven members usually become overly complex and more difficult to manage.

- Actively seek to discover strengths and weaknesses of the group as a whole and of members.

- Manage for conflict from the beginning. Accept that conflict is inevitable, but doesn't have to lead to problems. Identify sources of potential conflict and openly discuss them. Agree to disagree. (This is covered more in Chapter 17.)

- Openly discuss the group's goals so that everyone knows what is happening. Allow each person to openly express their personal viewpoint. This will uncover major areas of difference. Do not be judgmental on anyone's viewpoint. You are trying to establish commonalties and differences to help foster cohesiveness.

- Set up a schedule. Define the group's time constraints and set milestones. Identify when group members can meet, and what time restrictions exist. Set realistic goals and objectives for the group.

- Get agreement on leadership and facilitation. Who will lead or direct? Set up a known system for delegating tasks.

- Deal openly with issues of participation, permissiveness, and cohesiveness. These can be sources of conflict. Decide how members of the group will participate. If members don't follow through with designated tasks, decide how they will be sanctioned if necessary. Openly discuss what is or isn't allowed.

Work towards group cohesion with everyone being part of the team, and not just individuals working together. Is groupthink occuring (see Box 1)?

- Follow decision-making criteria

    — Solicit and review alternative courses of action. Review the complete range of objectives and the values implicated by a decision.

    — Closely examine the consequences of each alternative.

    — Search for new relevant information.

    — Incorporate all new relevant information.

    — Reexamine the consequences.

    — Make detailed plans for implementation and action of the selected final decision.

CAPACITY BUILDING

Effective organizations rely on strong, dedicated, and skilled individual leaders in order to achieve their goals. The development of effective organizations and individual leaders is capacity building. But, capacity building goes beyond merely producing strong individuals. It also involves getting groups to work toward common ends. So, groups working toward environmental goals can be linked through capacity building, as their leaders network and environmental efforts become more comprehensive and collaborative. Powerful coalitions can be formed among groups if they know of each other's existence and recognize their common positions.

Coalitions are powerful entities often formed to resolve situations that otherwise might be unresolvable. It encompasses using representatives from all identifiable viewpoints in a situation and ensuring every member of the coalition has equal status. The special factors used in understanding community groups all apply to coalitions. However, coalitions will most likely be formed of a larger and heterogeneous mixture of business/industry, agencies, and various community and special interest groups.

Environmental communicators can participate in capacity building through these activities:

Identify likely allies.

- Identify key people in key groups that share your group's agenda and role.
- Identify potential members and the reasons they may wish to join your group.
- Recognize potential barriers to forming coalitions. Develop strategies for dealing with inter-group conflicts.
- Develop strategies to make potential members aware of your group.

Involve all coalition members equitably.

- Establish a shared vision.
- Set realistic goals and objectives.
- Involve all members in some way.
- Evaluate the levels of cooperation among coalition members.
- Make meetings meaningful.

- Share successes among all.
- When conflict comes, reveal it, do not conceal it.

In conclusion, the forming of a coalition can lead to productive solutions. The communicator often acts as the facilitator or organizer and helps keep the coalition focused and on process.

FORMATS FOR PRESENTING INFORMATION TO GROUPS

As an environmental communicator, many of your messages will be crafted for and presented to groups. Recall that audience analysis purposely groups people. When your audience has already gathered itself together under a philosophical umbrella and given itself a label, as organizations have, your process of message planning and delivery should be better off. Also, as you are preparing to communicate with a group, it behooves you to find out what form of interaction your presentation can take. The communicator can take advantage of the kind of group by varying the format to meet the end goal. (More on presentations in Chapter 11.) Here are some examples of form and function of group presentation:

A Speech, Film or Demonstration

This gives information in an organized way but does not give audience a chance to talk. Such a presentation is most successful when delivered by a person who knows the subject thoroughly and can present it using visual aids. Combine questions and group discussion to get participation. This format allows a lot of information to be disseminated quickly.

Brain-Storming

This open-ended, non-judgmental technique gets many ideas out quickly. Members of the audience throw out an idea on the topic under discussion while one person writes them down on a board or easel. It is important to accept any idea, no matter how wild. A discussion of the ideas and how best to organize them then allows the group to see which ideas are worth pursuing further. This process can help produce a cohesive group and takes advantage of the "safety" of a group.

Buzz Sub-Groups or Small Discussion Sub-Groups

To use these sub-groups well, divide the group into sub-groups of 4 to 10, preferably around tables. Designate a discussion leader and discussion recorder for each group. Make the topic of discussion clear and focused. After allowing 5 to 15 minutes of discussion, bring the whole group together and get a summary report from each group. End with general discussion of all the points raised. This way everybody has the chance to take part in the discussion. This form is good for getting commitment to action, but hinders in-depth discussion.

Role Playing

To use role playing, give each participating member a card briefly explaining the character they will represent. Indicate clearly the points of view to be taken but not the exact words they are to say. Try to choose players who will take criticism in good spirit. Take the players aside and tell them to have fun with this exercise, but to remain in character during the discussion. Then, let them launch the role playing session. Stop the role playing before interest wanes and immediately follow-up with an open discussion of what just occurred. The main purpose here is to get people to adopt other ways of thinking. This technique helps in understanding attitudes and opinions of others. Role playing does not, however, always provide new information.

Panel Discussion

A moderator guiding three to five panel members with different views is a good technique for looking at various perspectives on a topic. Keep talk between the panel members informal. The moderator should not be a central figure in the discussion but should only invite comments and reiterate or ask questions if the discussion starts to become dry. Allowing a free flow of questions from the audience makes the discussion more inclusive and interesting.

The Colloquy or Talk-Show Format

This technique is similar to a panel discussion, with a bit more influence by the guide and more discussion from members of the audience encouraged. This is often used in free information exchange conferences, in early stages of coalition building, or issue resolution to identify community concerns.

The Symposium

In a symposium, each person on a panel of experts delivers a speech/presentation to the audience. A symposium chair introduces the speakers, provides transitions, and may provide occasional summaries. Usually this is a unidirectional communication mode and the audience needs to understand that its role is mostly limited to receiving information.

CONCLUSION

The concept of a group is important to the environmental communicator, because audiences most often consist of individuals who influence one another. Formal organizations, temporary project teams, and ad hoc community groups are just some of the groups that are common in resource management issues. Regardless of whether the communicator is a member of the group or communicating from outside the group, understanding the internal dynamics of the group are critical for understanding how members will transfer and understand messages.

Box 1: Groupthink

Sometimes group processes grind to a halt or seem to spiral downward. When efficiency and progress deteriorate, groupthink (Janis 1982) may have infected the dynamics between members. Groupthink is defined as "a pathological mode of thinking within groups where concurrence seeking becomes so dominant in a cohesive in-group that it tends to override realistic appraisal of alternative forms of action" (Janis 1982). In a case of groupthink, members tend to become too selective in the information used to make decisions. Alternatives are ignored. Facts and opinions contrary to the group's stated position are ignored, while facts and opinions supporting the groups position are readily accepted without question.

Symptoms of groupthink

- Believing the group is invulnerable.
- Rationalizing avoidance of warnings and threats.
- Believing the group is completely moral and just.
- Placing irrational pressure on member who expresses doubts.
- Members censoring their own doubts.
- Belief that members of the group are in unanimous agreement.

Prevention of groupthink

- Allow all members to be critical evaluators.
- Discuss all viable options.
- Assign one member to be the devil's advocate.
- Seek the opinion of knowledgeable but uncommitted outsiders.
- Separate into sub-groups to explore options.
- Think rationally.
- Assign one member to be a sounding board.
- Consider the consequences of and processes used to make prior decisions.

REFERENCES AND FURTHER READING

Adler, N. J. (1996). *International dimensions of organizational behavior* (3rd ed.). South-Western Publishers.

Beckhard, R., & Pritchard, W. (1992). *Changing the essence: The art of creating and leading fundamental change in organizations.* Jossey-Bass Publishers.

Conrad, C., & Poole, M. S. (1998). *Strategic organizational communication into the twenty-first century* (4th ed.). Harcourt Brace College Publishers.

Daniels, T. D., Spiker, B. M., & Papa, M. J. (1997). *Perspectives in organizational communication* (4th ed.). Brown & Benchmark.

Daft, R. L. (1997). *Organization theory and design.* South-Western Publishers

Forsyth, D. R. (1998). *Group dynamics*, 3rd ed. Brooks/Cole Publishing Co.

George, J. M., & Jones, G. R. (1999). *Understanding and managing organizational behavior* (2nd ed.). Addison-Wesley.

Gordon, J. R. (1998). *Organizational behavior: A diagnostic approach.* Prentice Hall.

Heimlich, J. E., & Norland, E. (1994). *Developing teaching style in adult education.* Jossey-Bass Publishers.

Janis, I. L. (1982). *Groupthink* (2nd ed.). Houghton Mifflin: Boston, MA.

Kennedy, D. (1998). *Breakthrough! Everything you need to start a solution revolution* (Book, CD-Rom, Problem Identification Card Pack). Leadership Solutions Publishing.

Kolb, D. A., Osland, J. S., & Rubin, I. M. (1995). *The organizational behavior reader* (6th ed.). Prentice Hall.

Moorehead, G., & Griffin, R. W. (1997). *Organizational behavior: Managing people and organizations* (5th ed.). Houghton Mifflin College.

Oyster, C. (1999). *Group dynamics.* McGraw-Hill.

Pearse, M., & Smith, J. (1990). *Community groups handbook* (2nd ed.). Journeyman/Community Development Foundation Publications.

Robbins, S. P. (1998). *Organizational behavior: Concepts, controversies, and applications* (8th ed.). Prentice Hall.

Schermerhorn, J. R., Jr., Hunt J. G. & Osborn, R. N. (1997). *Basic organizational behavior* (2nd ed.). John Wiley & Sons.

Skinner, S. (1997). *Building community strengths: A resource book on capacity building.* Community Development Foundation Publications.

Stewart, G. L., Manz, C. C., & Sims, H. P. (1998). *Team work and group dynamics.* John Wiley & Sons.

Tilden, F. (1957). *Interpreting our heritage.* University of North Carolina Press.

Turner, N. W. (1997). *Leading small groups: Basic skills for church and community organizations.* Judson Press.

Twelvetrees, A. (1991). *Community work* (2nd ed.). Macmillan.

11

DIFFERING WAYS OF THINKING AND DOING

For any form of communication to impart a message that triggers learning to occur within the recipient, the communicator must take into account the diverse ways people learn and make use of media. These different personality, learning and coping styles often may conflict when people work together. Personality styles affect people's interactions because of different ways that information is processed. Similarly, learning styles affect how people prefer to receive information. Richard Felder (1999) comments:

"[People] have different learning styles, characteristic strengths and preferences in the ways they take in and process information. Some [people] tend to focus on facts, data, and algorithms; others are more comfortable with theories and mathematical models. Some respond strongly to visual forms of information, like pictures, diagrams, and schematics; others get more from verbal forms—written and spoken explanations. Some prefer to learn actively and interactively; others function more introspectively and individually."

PERSONALITY TYPES

Learning is affected by personality and personality is heavily influenced by learning over one's lifetime. This interplay produces differences in how each individual prefers to process information and to deal with others. It is when we work in groups that personality types can conflict. Understanding different personality types should help everyone become tolerant of each other's differences.

Again, our intent here is to do no more than introduce these ways of categorizing personalities and lead you to instruments that will help develop your understanding of different personalities. A practical aspect of this for communications is to garner an understanding of how people interact, deal with problems, what they consider important, and how they look at the world. People are rarely static in how they react to the world. People will often react in different ways dependent on the situation in hand. But, most people have a preferred way, often unconsciously so, of reacting to situations and other people. Understanding why people may react the way they do reduces conflict.

Satir Modes

How do you view life? The following lists the Satir modes (Satir 1964; Elgin 1980) and describes how individuals communicate and make use of presupposition statements. Can you identify your mode? Note how the Satir modes can easily conflict with each other. Recognizing the different modes that may exist in a group will help in warding off unexpected conflict based solely on personality traits.

- **Blamer**—The type of person that best characterizes this mode feels alienated and is convinced that most people are oblivious to their needs and feelings. The blamer usually reacts with demonstrative dominance. They usually criticize others and externalize problems to other people. A blamer frequently uses statements such as "why do you do that," "you always do that," and "why spoil everything." Two blamers communicating tend to spread conflict readily.

- **Placater**—The person who typifies this mode is afraid of upsetting people and can go to great lengths to avoid conflict and alienating people. They are usually distant and anxious and frequently use statements like "oh, you know me, I don't care" and "whatever you do is fine by me." Often they do care, and it isn't fine, but they get worked up because no one has realized it. Two placaters communicating tend not to reach any consensus since neither will make a decision for fear of offending the other and hence losing support.

- **Leveler**—This person's *modus operandi* is to be open and tell it as it is! Sometimes, this occurs bluntly. Usually the verbal and non-verbal languages match. Two levelers tend to communicate well since each is being honest. But, if a person has adopted a leveler mode and is not honest, a conflict escalation may result because the true leveler is confused and becomes defensive because of the "phony" leveler.

- **Computer**—This is an emotionless state and also the mode in which non-verbals are at a minimum (like Mr. Spock or Data from the Star Trek series and films). If in doubt what to do or say, this mode offers the best option. There is less likelihood to misunderstand this mode or even a need to become defensive against it. This mode is less likely to lead to conflict situations; it will probably help to reduce conflict since the mode is usually logical and "to the point." Still, a lack of "outgoing cues" can make the recipient feel a little alienated. The person with a preference for the computer mode tends to fear expressing feelings and emotions. Statements from this type of person would include "one would think that. . ." or "it can be safely assumed that. . ." or "obviously, no cause for alarm."

- **Distracter**—This state emphasizes panic. A person in this mode continuously cycles through all the other four states. The non-verbals are also somewhat confusing since they often do not match the verbals and change frequently.

Myers-Briggs Types of Personalities

Myers-Briggs Type (Introduction to Type 1976) is perhaps the most widely used personality instrument. It was first designed by Carl Jung. Its scales are continuums and most of us will lay along them rather than be as polarized as the descriptions. In its essence, it uses four continuums of action and thinking. Like the Satir modes above, recognizing the different personality profiles helps in diffusing conflict before it occurs.

- **Extrovert to Introvert (E to I)**—This factor is not actually about being shy or outgoing. Rather, it is about how we focus our attention and draw energy from interpersonal contact. An extroverted person will usually focus their attention outward to people, thus engaging people in his or her thought processing. The introvert, however, tends to remain quiet and focus any thinking internally until they are ready to speak. Conflict arises when each type feels that the other is negating their thinking. Introverts mull problems over in their head until they feel they have a definitive answer and then will share it out loud. Extroverts, however, will probably speak out loud straight away and start running through options to resolve a problem. The extrovert is merely processing (thinking) aloud and not really negating the answer that the introvert gave, yet the introvert is more than likely to perceive that the extroverted person is not listening or just negating the decision. They may both arrive at the same decision, but the thinking preference is different. Knowing this helps resolve the seemingly silly conflicts that often arise from feeling that someone is "ignoring your worth."

- **Sensor to Intuitive (S to N)**—This is about how we acquire and use information. Sensors use direct factual information and their five senses to gather information for decision-making. Intuitives rely on creative imagination, inspiration, and intuition to arrive at a decision. Avoiding conflict here calls for sensors to trust and rely on the intuitives, and for the intuitives to accept that the sensors need sound reasoning to accept their arguments or decision.

- **Thinker to Feeler (T to F)**—This category deals with the rationale we use to make decisions. The thinker will use objective and logical information to make a decision—a sort of cost-benefit analysis. The feeler will use personal values to make a decision. This might imply that feelers can be dogmatic in their approach, refusing to budge on an issue that they feel will go against their values. Thinkers, however, might be seen as more willing to compromise in order to gain a consensual decision for the benefit of others, even though it may go against their personal values.

- **Judger to Perceiver (J to P)**—Are you planned and organized or spontaneous and flexible? People who are judging tend to need to know where every minute is going and how it is going to be spent before they do something. Perceiver tends to go wherever the winds of change take them. If you have ever been on vacation with a polarized judger, you will know just where you are going to stay and what you will probably eat, two weeks before you actually get there. The same vacation with a full perceiver will be an adventure, with you not knowing where you will be one hour hence.

Again, it is not only knowing your own preferences that helps you understand yourself, but rather understanding the opposite preferences. You can understand why you may be having conflict and then be able to work with those differences between yourself and others.

Enneagrams

Don Richard Riso discusses enneagrams which deal with the practical application of different types of personalities and how they interact. Enneagram types work on a continuum between a integrative (healthy) type or a disintegrative (destructive) type.

It should be emphasized that the following nine enneagrams are basic patterns, and individuals are unique variations of those patterns. Still, this model of personality gives an insight into the different types of personalities. It is critical that you not just pick out the profile that you would like to be! If, for example, without testing, you thought you were a type three because you see yourself as self-assured and ambitious, be aware that you also need to identify that you can also be narcissistic and psychopathic in that type. Each person has a basic type, and then can have two support wing types. The overall pattern dictates how a person is likely to view the world and react to situations that occur. Understanding both the positive and negative traits of each type, and especially how they can interact with your own, helps to reduce conflict situations based on different ways of thinking and doing.

In essence, the nine different types can be summarized as follows:

1. **Reformer:** Idealisitic and orderly (integrative) to perfectionistic and intolerant (disintegrative)

2. **Helper:** Concerned and helpful (integrative) to possessive and manipulative (disintegrative)

3. **Motivator:** Self-assured and ambitious (integrative) to narcissistic and psychopathic (disintegrative)

4. **Artist:** Creative and individualistic (integrative) to introverted and depressive (disintegrative)

5. **Thinker:** Perceptive and analytic (integrative) to delusional and paranoid (disintegrative)

6. **Loyalist:** Likable and dependent (integrative) to dogmatic and masochistic (disintegrative)

7. **Generalist:** Accomplished and extroverted (integrative) to excessive and manic (disintegrative)

8. **Leader:** Powerful and expansive (integrative) to dictatorial and destructive (disintegrative)

9. **Peacemaker:** Peaceful and reassuring (integrative) to passive and repressed (disintegrative)

As in the Myers-Briggs Types of personalities, it is essential to be able to understand yourself and others and how these different personality traits and preferences can conflict. Knowing these potential conflicts allows easy conflict management. It is also necessary to be honest with yourself when taking these tests. We all would love to be the perfect ideal, but it is in understanding your weaknesses that you really become strong.

LEARNING AND COPING PREFERENCES

In the literature, there are numerous different learning styles and even different intelligences. This chapter will just overview three of these concepts to indicate how pervasive they are and how the communicator needs to be aware of these different preferred ways of thinking, especially when doing personal presentations or workshops. Awareness of the differing ways of thinking and doing can help the communicator to reduce the noise associated with encoding and decoding the message. This is a form of multiculturalism, since different ethnic groups will often have ingrained cultural ways of thinking and learning. (Multiculturalism is covered more in Chapter 15.)

Field Dependent versus Field Independent

Some people can easily find a graphical line figure hidden in a maze of other lines. Others, no matter how hard they try, cannot even begin to find a hidden figure. It has nothing to do with intellect, but rather how we relate to the world. While most of us will fall along a continuum of this measure, we will discuss the extremes and how these types of people differ in their work and social thinking. Again, neither one is better or worse than the other—they are just different preferences in how they manage information and socialize. Those that have a hard time finding information buried in a maze of other information can be classified as "Field Dependent" or global thinkers, while those that seem to easily focus on a complex field of information and find discrete parts are "Field Independent" or analytical thinkers. Of course, those that fall in the middle will display traits from both ends of the continuum.

Characteristics of Field Dependent and Independent Thinkers

	Field Dependent	Field Independent
Relationship to peers	Likes to work with others to achieve a common goal Likes to assist others Is sensitive to feelings and opinions of others	Prefers to work independently Likes to compete and gain individual recognition Task oriented; is inattentive to social environment when working
Personal relationship to instructor	Openly expresses positive feelings for instructor Asks questions about instructor's tastes and personal experiences; seeks to become like instructor	Rarely seeks physical contact with instructor Formal; interactions with instructor are restricted to tasks at hand
Instructional relations to instructor	Seeks guidance and demonstration from instructor Seeks rewards which strengthen relationship with instructor Is highly motivated when working individually with instructor	Likes to try new tasks without instructor's help Impatient to begin tasks; likes to finish first Seeks nonsocial rewards

(continued)	Field Dependent	Field Independent
Characteristics of program that facilitate learning	Objectives and global aspects of program are carefully aligned Concepts are presented in humanized or story format Concepts are related to personal interests and experiences of learners	Details of concepts are emphasized; parts have meaning of their own Deals with math and science concepts Based on discovery approach
Content	1. Social abstractions: Field-dependent program is humanized through use of narration, humor, drama, and fantasy. Characterized by social words and human characteristics. Focuses on lives of persons who occupy central roles in the topic of study, such as history or scientific discovery. 2. Personalized: The ethnic background of learners, as well as their homes and neighborhoods, is reflected. The instructor is given the opportunity to express personal experiences and interests.	1. Math and science abstractions: Field independent program uses many graphs and formulas 2. Impersonal: Field-independent program focuses on events, places, and facts in social studies rather than personal histories.
Structure	1. Global: emphasis is on description of wholes and generalities; the overall view or general topic is presented first. The purpose of use of the concept or skill is clearly stated using practical examples. 2. Rules explicit: Rules and principles are salient. (Learners who prefer to learn in the field-dependent mode are more comfortable given rules than when asked to discover the underlying principles for themselves.) 3. Requires cooperation with others: The program is structured in such a way that learners work cooperatively with peers or with the instructor in a variety of activities.	1. Focus on details: The details of a concept are explored, followed by the global concept. 2. Discovery: Rules and principles are discovered from the study of details; the general is discovered from the understanding of the particulars. 3. Requires independent activity: The program requires learners to work individually, minimizing interaction with others.

As you read through the table above you may have noticed how differently the two ends of the continuum prefer to work and interact. Most people tend to use the mode with which they feel most comfortable. If other people are not comfortable with that mode, then strained interactions will probably result, or the audience will not be attentive.

Gardner's Multiple Intelligences

Gardner (1993, 1999) has developed a theory that we all have up to seven unique intellegences. We all have these intelligences, but some are more expressed than others. Gifted people may have a higher than average score on a specific intelligence yet score lower on a couple of others. Overall, most people score around average on a few of the intelligences with one or two higher or lower. These are listed as:

- **Spatial Intelligence:** People who are high in this ability tend to think in pictures and create vivid mental images to retain information. They enjoy dealing with visual stimuli and will probably work in positions that emphasize their visualization skills.

- **Linguistic Intelligence:** people who are good at languages have a high ability in this category. They can listen and speak well and tend to think using words. They probably work well in positions where speaking is a primary part of the work.

- **Mathematical Intelligence:** These people think in logical and numerical patterns, and are good at synthesizing information. They are also research-oriented and like to unravel problems.

- **Kinesthetic Intelligence:** A movement intelligence means people high in this category will prefer to express themselves through action by interacting with the space around them. They make good dancers and athletes, and are adept at physical control or manipulation of objects. Technicians who use sophisticated equipment will also be high in this category.

- **Musical/Rhythmic Intelligence:** People who are adept at music or can pick up a tune the first time they hear it are high in this category. They tend to be highly tuned to sounds around them and react accordingly. They tend to think in musical patterns.

- **Interpersonal Intelligence:** High empathy and an ability to easily view other perspectives emphasizes this category. They work to manage conflict. They are also highly intuitive and perceptive in nonverbal use. Counselors, motivational speakers, and good communicators will probably have this as a primary intelligence.

- **Intrapersonal Intelligence:** A person with a higher level of this intelligence will be highly self-reflective and be cognizant of the dynamics of relationships. They are most evaluative and aware of their own inner workings.

What is important to note here is that people who have a higher emphasis in one of the intelligences will prefer to use that mode in their communication. Being aware of your own modes and being sensitive to other modes will help when interacting to build groups and process information as a group.

Learning Styles

While multiple intelligences emphasize what modes we work best in, the following learning style emphasizes the kind of ways we prefer to learn. We describe the learning styles here. A hands-on test is available at the following web site: http://www.hcc.hawaii.edu/intranet/committees/FacDevCom/guidebk/teachtip/lernstyl.htm (Barsch/Haynie 1999).

- **Visual:** These learners prefer to see what is happening. Visual aids are most desired, and they need to have a full view of the speaker to garner nonverbal nuances. They will also take detailed notes to visualize and absorb the information.

- **Aural:** Aural learners prefer to hear the information. They are tuned into the speaker's voice patterns and inflections. Audio-books are an ideal medium for these learners.

- **Print:** Learners with this preference prefer to read their information from text and usually alone.

- **Kinesthetic:** These people are usually fidgety and need to move around. Sitting still bores and distracts them. When speaking they are best when allowed to pace slowly or when involved with equipment. As learners they need to have something physical to do. Usually an animated group that loves to explore.

- **Haptic:** People in this category prefer to feel objects and work hands-on. They learn through doing and touching. This group is similar to the kinesthetics, except they have more patience and the high energy is not usually present.

- **Olfactory:** This category prefers odors and smells to recall information and trigger the memory. Obviously a smaller group than the others, but a powerful way of remembering information.

Note how people with different learning preferences will prefer different modes of interaction. While some will be willing to sit and listen to a speaker, others need to be doing something active in order to be attentive. Assessing your audience and varying your presentation techniques will result in a more attentive audience.

Many more learning style instruments, and references to others can be found in Bennett (1998). Alternatively, a web site with more information and links for other ready-to-use style tests can be found at: http://snow.utoronto.ca/Learn2/resources.html

CONCLUSION

A lot of research has gone into understanding how people think and learn. If you ever asked any of the following questions during meetings as a participant listening or interacting, or as a communicator, then you might now have an insight into:

- Why are the meetings sometimes fraught with tension and misunderstandings? And, yet at other times they flow well, even when I do the same technique. What's up with that?

- Why are some of participants so frustrated, even hostile?

- Why does the communicator assume that everyone attending this session hears and understands the information with such a loud ventilation system?

- Why don't the audience understand the "simple" schematic diagrams I'm giving them?

- Why do the audience want a visual when I explained it clearly enough?

- Why is the audience so unresponsive even though the lecture was well-planned?

- Why don't those people get it, when it has been explained so clearly?

Obviously, the answers to these questions are rooted in how people prefer to learn and how they cope with different situations. Individuals vary considerably, but many cultures seem to show predominant traits in particular styles of learning, thinking, or acting. Much of this information is more applicable in face-to-face activities such as personal presentations or when helping communities during educational interventions. It is a little more limited when using the mass media. What factors may come into play that will stop the audience from comprehending the message? Different cultural factors will be one aspect. Different learning, thinking or coping styles will be another. Dependence on the audience size, location, and mode of communicator interaction will affect how the message needs to be developed and disseminated.

REFERENCES AND FURTHER READING

Banks, J. A. (1998). *An introduction to multicultural education* (2nd ed.). Allyn & Bacon.

Barsch, J. (1999). *Learning style test*. Adapted from Barsch Learning Style Inventory, and Haynie, Nancy A. Sensory Modality Checklist. http://www.hcc.hawaii.edu/intranet/committees/FacDevCom/guidebk/teachtip/lernstyl.htm.

Bennett, C. I. (1998). *Comprehensive multicultural education: Theory and practice* (4th ed.). Allyn & Bacon.

Blanchard, K. H., Carlos, J. P., & Randolph, A. (1999). *The 3 keys to empowerment: Release the power within people for astonishing results.* Berrett-Koehler Publisher's.

Elgin, S. H. (1980). *The gentle art of verbal self defense.* Prentice Hall.

Felder, R. (1999). Matters of style. http://www.asee.org/pubs/html/styles.htm

Gardener, H. (1993). *Frames of mind: The theory of multiple intelligences* (10th ed.). Basic Books.

Gardener, H. (1999). *Intelligence reframed: Multiple intelligences for the 21st century.* Basic Books.

Goldberg, M. J. (1999). *The 9 ways of working: How to use the enneagram to discover your natural strengths and work more effectively.* Marlowe & Co.

Introduction to type (2nd ed.). (1976). Center for Applications of Psychological Type: Gainesville, FL.

Riso, D. R. (1990). *Understanding the enneagram: The practical guide to personality types.* Houghton Mifflin Co.

Satir, V. (1964). *Conjoint family therapy.* Science and Behavior Books, Inc.

Satir, V. (1972). *Peoplemaking.* Science and Behavior Books, Inc.

12

COMMUNICATING ACROSS CULTURES

"Culture" is one of those expansive terms that we use most always without an explicit definition. Like "freedom," "love," and "common sense," culture seems to be intuitively understandable. But, grasping its meaning gets mentally slippery when we attempt to saddle it with a cut-and-dry definition. Ideas contained in the concept of "culture" include:

- integration of a society's knowledge, beliefs, and behaviors;
- education so as to transmit what is important to new members of a nation;
- customary beliefs, social norms, and material artifacts of a group;
- the body of shared beliefs, values, attitudes, opinions, and practices of an organization.

Imparting culture can be thought of as the same as education, though then we are using another concept that is larger than any definition. Undergirding the upkeep of culture is communication. The integration and transmission of the intangibles that make up society must involve messages. Lots of them.

So, how can we consider culture and use any insights we gain to communicate better? Clearly, communicators deal in big ideas. Lots of them. Thinking back to the communication model in Chapter 2, we need to consider the concepts of encoding and decoding of information. Even when an audience has been well characterized, there still exists the "noise" that occurs because the audience cannot clearly decode the message. Multicultural aspects can cause receivers of the message to perceive things differently than was intended by the communicator.

Culture and communication are interdependent. One cannot exist without the other. Culture provides the blueprint that determines the way an individual thinks, feels, and behaves in society. Communication gets those points across. We are not born with culture engrained in us, but learn it through acculturation and socialization. It is manifested through societal institutions, daily habits of living, and fulfillment of psychological needs. Breach a cultural norm and those around censure you. Do well by their standards and they reward you with praise and admiration.

For environmental communicators, cultures can be conceptualized as analogous to ecosystems in the way they are nested inside one another. Drawing boundaries around either cultures or ecosystems may be necessary but is also ultimately imperfect. Environmental communicators, who carry with them intimate understandings of both the human world of culture and the wider world of ecosystems, are unusually suited for linking these two worlds and for passing that linkage onto others. This quality of seeing the human world within the larger natural one is environmental sensitivity. Environmental sensitivity has been shown to be one of the best indicators of environmentally responsible behavior (Sia et al. 1985/1986).

An awareness that there is no monolithic culture and that individuals come from varied and diverse cultural backgrounds is essential if a focused message, through whatever mechanism, is to be delivered successfully. A helpful concept to keep in mind when considering the broad spectrum from which members of your audience come is microculture—those cultures within cultures within even broader macrocultures in which we all exist.

These varied microcultural structures are emphasized in figures 14a and 14b.

Understanding why another person is reacting differently to a situation than you is a wonderful way to reduce conflict and to smooth differences, and also to garner understanding of a reality other than your own. We all have our own realities and worldviews. Until we are exposed to other realities we may not be able to understand why some people cannot see our points of view. Many people conflict, not on content, but on the process of communication itself. Understanding some of the different ways people process information and work through situations will help smooth communication.

It should be emphasized that our early development defines who we are and what we accept. In order to understand others, we need to see outside of our own personal boxes of reality. Exposing yourself to disconfirming information and other ways of being is a wonderful way to truly see your own particular set of cultures and to begin seeing other realities.

We all live and operate in many microcultures. But, only an individual who can successfully operate in two or more macrocultures can be termed bicultural or multicultural. Operating means to be able to work, understand, and communicate easily within the culture where one is. The last decade has seen rising value placed in multiculturalism. This is, perhaps, one corollary of the global village effect that became more and more evident in the last 50 years. Communicators are expected to be sensitive to the spectrum of cultural backgrounds of their message recipients.

MULTICULTURAL MODELS

Macro versus Micro

Macroculture (also known as dominant culture) is the national culture that is shared by most of a nation's residents. In addition to participating in the macroculture, each individual also belongs to a number of other cultures with patterns that may not be common to the macroculture. Using the United States as an example, within the dominant American culture one can easily recognize regional, ethnic, and socio-economic variations that result in cultural variation. Think of Southern culture with its cuisine of sweetened iced tea, fried chicken, and grits. Or, Hispanic/Latin culture with its low-riders, burrito wagons, and rhythmic music. Compare those with Midwestern agrarian culture: the culture of hot dishes, working by the seasons, and family reunions where no one has to cross a county line to attend. And, the upper crust blue bloods in eastern suburbs, with their inherited wealth, country club memberships, and immaculate mansions. America has been called a melting pot, though many argue it is more like a tossed salad. Cultural differences within the macroculture make either of those metaphors accurate. An individual belongs to many different microcultures.

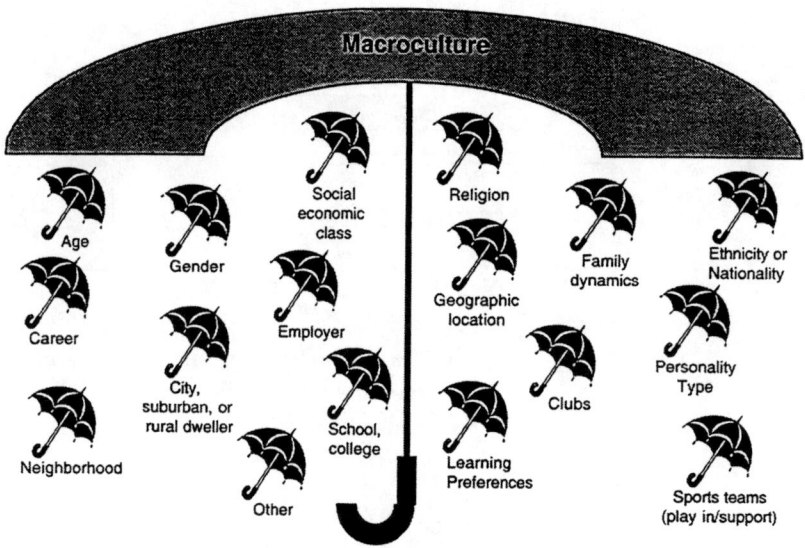

Figure 12a   Cultural Identity Umbrella Model
The many different umbrellas under the macro-umbrella represent
memberships in various groups and organizations that mold the
individual.

All those life ways in which you participate locally and regularly are your microcultures. Cultural identity occurs through the beliefs, traits, and values learned through membership in microcultures based on national/ethnic origins, religious affiliation, gender, age, socioeconomic level, primary language/dialect, geographic region, place of residence (e.g., rural, suburban, urban), and other exceptional and non-exceptional factors. The interaction of various microcultures within the larger domain of macroculture determines an individual's cultural identity. Membership in one microculture usually influences characteristics, beliefs, and values of membership in other microcultures. Therefore, an individual is unique, but can still be defined by some characteristics associated with a specific microculture to which she may be a member. The importance of each microculture will vary between individuals even if they share similar memberships.

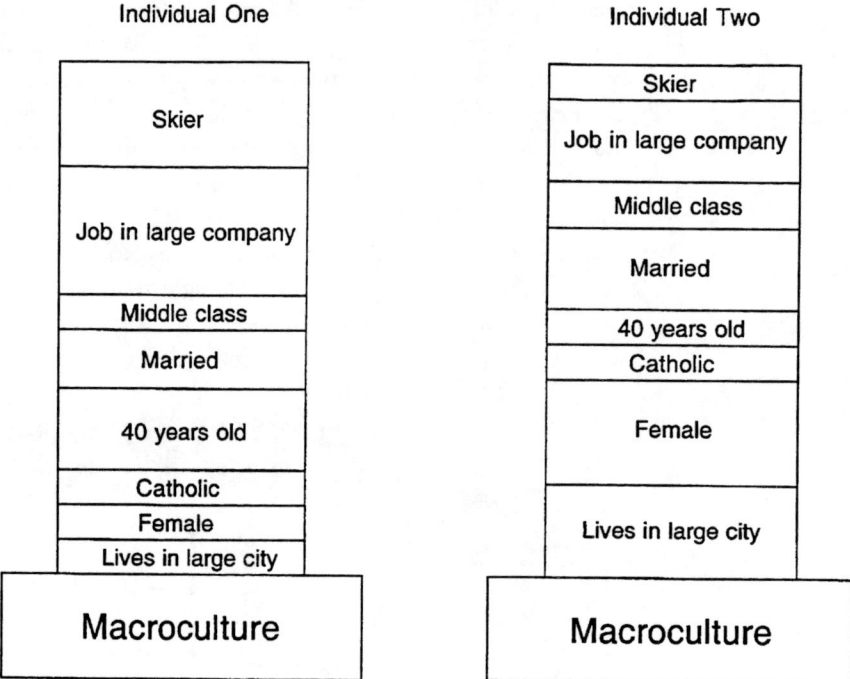

Figure 12b   Macroculture vs. Microculture Model
*Note that even though the two individuals have similar microculture
identities, the importance of each is different. Macroculture is the
foundation upon which the microcultures rest, and microcultures are
the many blocks in the column.*

CULTURAL ADAPTATION THEORIES

Individuals are exposed, literally from the moment of their birth, to the details
and telling aspects of their particular cultural milieu. This does not mean that there
is no choice and role of free agency in this cultural education. But, to learn a culture one must have some exposure to it.

The process of learning cultures has been looked at and theorized on by hundreds of scholars. We'd like to look at several of the more accepted theories of cultural adaptation. It is not our purpose here to debate the strengths and weaknesses
of any of these theories and especially how they might be, or might have been,
applied to individual cases. Rather, we merely want to introduce them, so that you
have some tools for looking at "different" groups.

Assimilation—This is a process in which microcultures may be adopted into
the macroculture. The process behind this concept is that there should be no
competing dominant culture outside of the macroculture.

- **Conformity theory**—this concept assumes that the basic tenet of the dominant society is to encourage adoption (forcibly or voluntarily) of the macroculture to the exclusion of all microcultures.

- **Melting pot theory**—emphasizes evolution of the macroculture through a
mixing of all culturally unique groups, to form a new culture expressing traits
from all the cultures.

Suppression—In a suppressive situation, a microculture is deliberately isolated from the rest of the macroculture (by force or by choice). While the people in this isolated culture may develop a "dualism" of identity, they are often viewed by the rest of the macroculture as inferior, and are often tightly controlled and/or sanctioned by all aspects of the society they live within.

Pluralism—This theory stipulates that cultural groups, particularly ethnic groups, maintain separate and distinctive identities from the dominant culture or macroculture. This is the idealized multicultural situation which operates through mutual respect and equity of all cultural groups.

Using these ideas to analyze and understand an audience informs message-creation for the communicator who takes time to inspect cultural nuances. Beginning to know the different microcultures from their particular perspective is helpful in forming an effective message that will be heard, understood, and acted on. One has only to think of the many minority groups in the United States, or any other county over the last two centuries, to begin realizing how these theories have all been applied at some time or other to some microculture.

WORLDVIEWS

A worldview is usually associated with an individual's concept of reality, but is also integral with a culture's view of itself. This is particularly true of how cultures relate to the natural world. Worldviews are derived from the cultural beliefs and values that a society forms over long periods of history. It is these factors that are transferred from generation to generation and form the fabric of how people perceive their environment and determine what is supposed to be important. While worldviews may change slowly over time, understanding why a culture has developed specific traits of behavior will aid the communicator in formulating a message that will be decoded accurately as it was encoded.

It is often noted that certain world populations and cultures share many characteristics. Sometimes, being aware of a worldview can be the key between a successful audience analysis and communication, and a failed communication plan. Many worldviews can be equated with Hall's concept of culture that sets cultures into High context and Low context (Hall 1976). Low context cultures tend to rely primarily on verbal messages with low emphasis on non-verbals. High context cultures have the opposite emphasis. The communicator should understand what type of cultural context is prevalent with any group that is being targeted with a message. Knowing the cultural context will help focus the quality of the message for a given targeted audience based on knowing what context is important for that audience.

The table on the following page, condensed and adapted from Bennett (1990), summarizes this concept:

Worldviews from a High and Low Context Perspective

	High Context	Low Context
Time	Polychronic (time not critical):	Monochronic (time of essense):
	Abstract view of time; Loose schedules; Lots of activity; Quick changes	Concrete need of time; Tight schedules; Linear use of time; Stick to times
Space & Tempo	High-sync:	Low-sync:
	Valued harmony with others and nature	Isolation from others and nature acceptable
Reasoning	Comprehensive logic:	Linear logic:
	Affective value; Intuition, spiral logic, and contemplation of value	Cognitive value; Logical, analytical reasoning of value
Verbal Messages	Restrictive codes:	Elaborate codes:
	Reliance on nonverbal and contextual cues with "shorthand speech." In-depth, interpersonal, socially cohesive, and polite communication more important.	Extended understanding of what is said or written with little reliance on cues. Little personalization with emphasis on argumentation and getting to the point.
Social Roles	Tight structure:	Loose structure:
	Behavior predictable; Role conformity expected	Behavior unpredictable; Role conformity context based
Interpersonal Relations	Group is first:	Individual is first:
	Well defined and accepted status distinctions. More in-depth bonding and highly structured inter-relational dependency by all members of the group. Differentiation and/or mistrust of outsiders.	Less defined status distinctions. Functional interrelationships within the group with little co-dependency. Less differentiation between members of group and outsiders.
Social Organization	Personalized law and authority:	Procedural law and authority:
	Customs and personal contacts are most important. Oral agreements are binding. When ordinary paths of action fail, contacts can "bend" rules to affect action. Authority figures are completely responsible and liable for all actions by subordinates.	Procedures, laws and rules are most important. Written contracts are binding. Impersonal, often unyielding policy rules all paths of action. Authority figures negate responsibility when consequences are negative.

Examples of how three different world cultures might differ through worldviews are (Hancock 1999):

African Worldview:

• Unity—Everything is functionally connected

- Oral tradition—oral history is important; language is participatory
- Survival of the group/Self-concept—individual finds identity in the "We-ness" of the group; the group is primary
- Extended kinship—each person is related to all other members of the tribe, ancestors, unborn, etc.
- Time—measured experientially; a focus on past and present; relaxed time
- Perception of the environment—important to be in harmony with the environment
- Activity—everything is continuous and connected; a sinusoidal view of activity

Asian Worldview:

- Hierarchical authority—favorable view of authority; respect age, social status, teachers, parents, etc.
- Filial duty—respect and loyalty for authority figures is important
- Collectivity—The group is more important than the individual
- Self-concept—identity of the individual evolves through group membership, family, microculture, etc.
- "*Gaman*"—value is achieved through suffering and hard work
- Social conformism—obedience to rules and regulations is highly valued
- Activity—everything works together in harmony; a helical view of activity

Euro-American Worldview:

- Competitiveness—be the best you can be
- Time—future oriented
- Belief in work ethic—value achievements and completion of tasks; visible and materialistic possessions define worth; delayed gratification
- Individualism—values individual effort over group effort
- Dualism—body and mind held as separate; concrete concept of right and wrong
- Perception of the environment—to be mastered and controlled
- Self perception—separate from the physical world; autonomy encouraged; solve one's own problems
- Activity—isolate a problem, solve it, and get on with the next problem; a stepwise progression of activity

Different microcultures derived from different ethnic origins will tend to incorporate the worldview of the original parent country e.g., Latin Americans, African Americans, Italian Americans. Others will have kept their original worldviews even under the overshadowing macroculture e.g., Native Americans. The worldview not only determines how these cultures look at the world but how they think and even feel.

Stereotyping versus Sociotyping

A term often used when talking about cultures is stereotyping. It is often used in a deprecating way. But, communicators must characterize their audiences to successfully target a message. Here we define the terms of stereotyping and sociotyping and differentiate between them.

Stereotyping

Psychologists have demonstrated that categorization by group, or stereotyping, is a natural phenomenon that humans use to develop mental categories that help make sense of the environment. We all stereotype because it is not feasible for our brains to process all the information continually bombarding us from our environment without some form of mental cataloguing. There is also a natural tendency for each of us to simplify our problems and to solve them with as much cognitive ease as possible. To analyze the behaviors, beliefs, and values of every individual with whom we might interact would be an extensive time consuming process (e.g., Bruner et al. 1986).

Stereotyping helps us develop a sense of self through our connection with socialized groups. It helps us define who we are in relation to the rest of the world and in turn develops our specific worldview. Such categorization of information also helps us in relating to our environment. When we come across certain symbols, behaviors, and patterns of speech, we expect a whole series of behaviors to occur, based on past experiences. These expectations then determine our reaction to a situation without having to think about it all that consciously. The broader the categories we learn to use, though, the less likely our overall perceptions are to be correct. These mental categories can often become too simple and cause us have a distorted perception of the world. But, it should be noted that this is what creates our worldview (Triandis 1971).

Stereotyping itself is not really either positive or negative. But, when stereotypes are overly broad or used without occasionally checking their veracity, then it can be dangerous. Such misuse may cause misconceptions, exaggerations, and inaccurate generalizations to be developed about the environment and those groups in it. Stereotypes used to describe all members of a group without the possibility of exception is prejudice, and this is at the heart of history's never-ending supply of ethnic and international tensions.

Subtyping

Many times we will have a fixed stereotype of a group, and then we will meet an exception that doesn't fit "the mold." Exceptions to a stereotype which are encountered are usually subtyped. This means that they are categorized apart from the main stereotype, but still associated with the stereotype as a deviant member of that category. This seems to be a cognitive mechanism to help us maintain the original (and probably inaccurate) category. If, however, we come across enough exceptions, we may begin to change that stereotype. Stereotypes are resistant to change and usually only do so with much conscious thinking and over a long time. This process is not without cognitive dissonance. It is best to recognize that we do stereotype and then to consciously make no judgment about the categorization we form. In making no judgments we can then see the group as a collection of unique individuals sharing some common theme.

Sociotyping

We have just given that categorizing is a natural phenomenon. In audience analysis, it is necessary to define a group so that a message may be correctly focused with the needs of the sender and the interests of the receiver. A sociotype is another form of categorization that is more short-lived than a stereotype. It is usually a more accurate generalization about a social group or microculture. Rather than placing a value on individuals having membership within a group, sociotyping just describes the characteristics common to a group. In a sociotype, an assumption is made between the relationship of a specific group and a chosen attribute. But, not everyone in the group is assumed to exhibit this attribute. Sociotyping is therefore concerned with what characteristics individuals share within a defined group, and not how that group defines an individual. A sociotype is a conclusion reached by a communicator about an audience after analyzing it. Once the message for that audience is transmitted, the sociotype is allowed to dissipate. It no longer is needed. By not seeking permanence, sociotyping avoids the dangers of stereotyping.

Special Sensitivity for People with Disabilities

When interacting with people with disabilities, the sensitive communicator recognizes their needs, without being condescending or paternal. "The challenge is to recognize that [disabled people] are more like [others] than they are different; to treat each as an individual; and to help them all achieve the greatest possible independence and personal, social, and academic development (Weisenstein 1986)."

To share messages with people with disabilities, you need to:

- Treat them equitably.
- Analyze the situation and their needs.
- Select the most appropriate media and instructional technique for the situation.
- Adapt the technique to make it most suitable for what you are trying to do.
- Be patient and tolerant.
- In group sessions, do not embarrass them by calling attention to their disabilities.
- Do not fill in words for those who stutter.

Sensitivity should be shown to all people, but some special empathy may be given to people with disabilities. The key word here is empathy and not sympathy. People with disabilities usually have well-defined goals and understand their limitations (Weisenstein 1986).

CONCLUSION

What is at the heart of this chapter is the need to be tolerant and understanding of differences we all have. These differences are based on cultural and social groups in which we are members. Depending on the culture into which we were born, we will have certain ways of behaving and also expectations about how things should be done. While categorizing is a normal part of thinking, we need to resist placing values on those categories. For the communicator developing a message it is paramount to understand the nuances of a culture. This will ensure correct encoding of a message that not only will be decoded correctly and understood, but will not offend the target audience.

However, with using the mass media, special emphasis should be given to cultural sensitivity. For instance, there are many gender-biased terms in use today. Terms like "mankind" can be easily replaced by "humans," or "people." Likewise, when referring to different ethnic cultures, terms that would appear derogatory must be identified and substitute terms used. Insulting terms are a form of noise that make your audience turn away from your message. The encoding and decoding aspect of the communication model should be foremost in the communicators' mind whenever they produce a message.

REFERENCES AND FURTHER READING

Adler, N. J. (1996). *International dimensions of organizational behavior* (3rd ed.). South-Western Publishers.

Banks, J. A. (1998). *An introduction to multicultural education* (2nd ed.). Allyn & Bacon.

Bennett, C. I. (1990). *Comprehensive multicultural education: Theory and practice* (2nd ed.), pp. 55–56. Allyn & Bacon.

Bennett, C. I. (1998). *Comprehensive multicultural education: Theory and practice* (4th ed.). Allyn & Bacon.

Bruner, J. S., Goodnow, J., & Austin, G. A. (1986). *A study of thinking.* John Wiley & Sons.

Gollnick, D. M. & Chinn, P. C. (1997). *Multicultural education in a pluralistic society* (5th ed.). Prentice Hall.

Hancock, C. (1999). Personal notes on worldviews. The Ohio State University.

Kenton , S. B., & Valentine, D. (1996). *Crosstalk: Communicating in a multicultural workplace.* Prentice Hall.

Lynch, J. (1986). *Multicultural education: Principles and practice,* Routledge & Kegan.

Lynch, J. (1989). *Multicultural education in a global society.* Taylor & Francis.

Schultz, F. (Ed.). (1999). *Multicultural education 99/00 (multicultural education 1999–2000)* (6th ed.). McGraw Hill College Division.

Sia, A. P., Hungerford, H. R., & Tomera, A. N. (1985/1986). Selected Predictors of Responsible Environmental Analysis: An Analysis. *Journal of Environmental Education,* 17(2), 31-40.

Seelye-James, A., Seelye, N., & Knudsen, A. (Eds.). (1994). *Culture clash: Managing in a multicultural world.* NTC Publishing Group.

Tayeb, M. H. (1998). *The management of a multicultural workforce.* John Wiley & Son Ltd.

Triandis, H. C. (1971). *Attitude and attitude change.* John Wiley & Sons.

Urech, E. (1998). *Speaking globally: Effective presentations across international and cultural boundaries.* Kogan Page Ltd.

Weisenstein, G. R., & Pelz, R. (1986). *Administrator's desk reference on special education.* Aspen Publications.

13

SPEAKING TO AN AUDIENCE

Most jobs today require both the ability to write cogently and to speak well. While writing skills are practiced from our earliest grade school, public speaking usually gets only scant attention in our academic development. This chapter is designed to help you become a polished speaker. Specifically, three aspects of public speaking will be covered: structuring the presentation, delivering the presentation, and overcoming anxiety when speaking in front of an audience.

STRUCTURING THE PRESENTATION

A superior presentation is well-structured, with an introduction, main body, and conclusion. Main ideas are then more likely to be received and remembered by the audience. Just like a written document, a good presentation is created by developing an outline and working through several drafts. Preparation and practice help ensure the presentation flows well, and help to reduce anxiety many of us feel speaking in front of groups. Even if you know your subject intimately, the presentation must be developed for the specific audience in order to be effective. As always, know who the audience is and why this group is there to listen to you. How much time do you have to present? What do you want to achieve with this presentation? Where will you speak? Are there suitable facilities for what you wish to do? Being prepared and knowing the audience, room and setting helps reduce stress and helps you to become comfortable with speaking before people.

Your Presentation's Introduction

The introduction should grab the audience's attention and orient them to your subject. Some ways to gain their attention are listed below:

- Ask a question of the audience (to be answered in the presentation).

- Refer to specific people in the audience (assumes some familiarity with the audience).
- Make a reference to a recent event familiar to the audience.
- Use an illustration or tell a humorous or dramatic story to set the scene.
- Use audio-visual aids—cartoons work well for this.

In the introduction, the main theme of the presentation should be stated, along with the main propositions that support the thesis. Often, the introduction includes an explicit statement of the goals and objectives the speaker is trying to achieve. Many speakers use feedforward to overview the talk and build interest at the same time.

Finally, the introduction should set the stage for you, as the center of attention. If a previous speaker introduced you, thank and acknowledge this, otherwise introduce yourself. Let the audience know how long you will be speaking and whether you want questions during the talk so that the audience knows what is expected of them. Many speakers, because of nervousness, apologize to the audience for lack of preparation or lack of knowledge of the subject. Remember that if you have prepared, this is not necessary and only makes everyone more nervous.

Your Presentation's Main Body

The body of your presentation should consist of three to five points that support the thesis of the talk. Effective speakers do not read the text of the presentation, nor do they memorize the body of the presentation. You should know the outline of the body well enough that you can "hold a conversation" with the audience while still following the outline. This will create a relaxed, natural presentation rather than a speech which sounds mechanical or phony.

Develop an expanded outline of the presentation to organize your thoughts. Then create a summarized outline sheet as a guide to organizing yourself and the visual aids you might use. If you cannot use visual aids, then keep your presentation outline simple with just a word or two to trigger your thoughts. It is important to keep it simple so that you don't have to hunt on the sheet for your next continuation, and also so you will not be tempted to read off your notes. If you can, develop visual aids as your primary guide. Each visual should prompt you on what to talk about next.

Your Presentation's Conclusion

The conclusion should summarize the points of the presentation, and provide closure for the audience. The summary includes a restatement of the thesis and/or objectives of the talk, the supporting points from the body of the presentation, and it indicates the importance of these points. Closure can be achieved by choosing an interesting quote that summarizes the thesis, by referring to events to follow the presentation, by providing a challenge to the audience, or by referring back to the introduction. In the conclusion, the speaker should not add new material, apologize, or keep talking. The conclusion should be concise and clean. A polished introduction and conclusion will set the audience at ease and create a professional image for the speaker.

DELIVERING THE PRESENTATION

Verbal Delivery

Verbal delivery is the characteristic we notice most in outstanding speakers. While delivery styles will vary according to the topic and the individual speaker, some general rules apply:

- Focus on the subject matter and not yourself. Do not wander off onto other subjects. The biggest problem we have when speaking is to constantly evaluate our own performances which is nearly always much more critical than anyone who is viewing us in the audience. The audience is there to hear what you have to say and, believe it or not, they actually want you to succeed.

- Practice in front of a mirror. If you are not looking at yourself most of the time, either because you are looking at your notes or because you are not making eye contact, then your style needs some work.

- Do not depend on reading a script. Instead use key words from an outline. It will make you look more professional.

- Know the material and sound confident (even if you don't initially feel so). By projecting confidence, you also project credibility. You will begin to feel more comfortable in front of the audience.

- Don't bluff when answering questions. If you don't know the answer, say so.

- When the subject has been covered, stop. Do not ramble on too long. If you have a time limit, adhere to it. Once you have talked over time, the audience is probably no longer listening and instead will be anticipating your conclusion. Don't dissappoint them.

- Be polite, smile, and display respect for the audience. It helps the audience and you relax.

- Make eye contact with the audience. Let them know that you are including them in your delivery. If they feel that you are talking with them, they are more apt to listen.

- Show passion for your topic. Enthusiasm will more than make up for many of the discrepancies you may exhibit in a presentation.

- Be aware of your time limitations. Be aware of your time limitations. Be aware of your time limitations.

- Two golden rules of public speaking:

  i. The talk should be well thought out and prepared.

  ii. Be realistic: there is no such thing as the "perfect speech."

Ethos relates to the speaker's credibility and the concept of "ethics." It covers the whole essence of what character you are trying to project. As a speaker, you have to maintain credibility for your message to be conveyed to your audience. This includes exuding a sense of confidence, knowledge about your subject, and empathy for the topic. Having an enthusiastic and sincere delivery keeps your audience interested. Finally you should look professional.

The speaker's appearance is an important factor in a successful presentation. Any presenter should want to appear knowledgeable, credible, trustworthy, and acceptable to the audience. As a rule, dress in clean and neat attire. This affirms your status as a serious professional. It can be a judgment call, but to dress casually can make you feel more friendly. Yet, it could also depreciate your credibility as a professional. Most people have a stereotypical image of what a knowledgeable professional should look like. Correct dress codes may appear "old-fashioned" but that is what a majority of people are concerned with—old-fashioned values. Looking as though you just stepped out of a week living wild in the woods might be functional for a specific interpretational talk in a campground, but otherwise might lead your audience to question your credibility.

Vocal Qualities

There are some basic qualities that define a well-projected voice. Even if one does not have the wonderful natural qualities like actor James Earl Jones, you can still be a dynamic speaker who captures the attention of the audience. Aspects that are crucial to being a well-rounded and credible speaker are inflection and projection, which are determined by pitch and loudness.

Inflection is determined by the pitch (tones of higher emphasis) and pace (delivery rate) at which we speak. While we all will agree that a person who drones on in just one predominant tone (monotone) at a constant pace can be boring to listen to, we must also be aware that the use of tones that become cyclic can also become boring to the listener after a few minutes.

In American and many British "English" dialects, there are usually only three to four tones, or sometimes five tones (pitch levels) that are used to emphasize semantic meaning within the language. A practiced speaker will try to vary the "rhythm" of the tones, slowing and speeding the pace of delivery a little, to emphasize the more important or critical points within the content. It is also useful to clearly emphasize key words. Low tones can be used to indicate negative connotations, and higher tones to emphasize positive aspects and extra emphasis.

Think of using the terms "good dog" and "bad dog." A dog only hears the tones. If "good dog" was spoken with a lower tone at start and finish, the dog might not be too happy to see you. Similarly, if both words in "bad dog" were said in higher tones, the dog would think it was being praised. It is usual to drop to a lower tone at the end of sentences to indicate the end of a spoken clause. But, it can be more effective to use a lower tone at the end to indicate something negative, or a higher tone to indicate something positive. Read, without emphasis, the following sentences, "It was announced that 50 American Elm trees had died from the blight. However, 200 more had recovered because of the new treatment." Put a lower tone on "blight" and a higher emphasis on "treatment" and notice how differently the sentences become.

Projection deals with the loudness at which we speak. We all know the difference between a whisper and a shout, but we can have difficulty making our voices heard in a larger room without shouting. In a small group meeting it is fine to speak with regular conversation volume, but in a larger room or an auditorium, especially one without a microphone amplifier, it can become almost impossible to hear someone who is not projecting their voice. Shouting is not a viable choice, because it will strain the vocal chords and may deafen the people up-close. Projection, or carrying power, eliminates the need for harsh or shrill vocals. In order to avoid strain on your vocal chords, you should determine your "optimum pitch" before any presentation. This can vary slightly at different times of the day and even on different days, dependent on mood and fatigue, but it should be within one or two tones of your habitual pitch. Habitual and optimal pitches are determined as follows:

- Read aloud a piece of text without any inflection, until you can maintain a monotone. This is your habitual pitch which you use in everyday conversation.

- Speak down the scale and find the lowest tone you can produce. Next, count the tones as you speak up through the scales until you reach the top note you can squeak out. Your optimum pitch is about the fourth or fifth from the bottom and will feel the easiest one with which to speak in a regular voice while projecting loudest. If your optimal pitch is not close to your habitual pitch, then practice speaking until the two are closer. This will help you to reduce voice strain.

In any setting as a speaker, you need to speak so that the people farthest away from you can hear easily and clearly. Breathing correctly is another important step to projecting well. It should come through the lower trunk of the body. While breathing, firm up your stomach muscles. This is called abdominal-diaphragmatic breathing. Try to push your stomach out while you breathe. You might want to try and push your stomach out against your hands until you get used to this method of breathing. Practicing this will allow you to control your exhaled breathing through the control of your diaphragm. Now couple this controlled breathing with well-formed vowels, and clear and precise articulation while you speak. Learning to relax the jaw will also yield fuller, firmer and resonating tones.

As a way to practice and actually see your breathing power during projection, see how you can affect a candle while breathing out using diaphragmatic control. (Fetzer 1984). Pick a short phrase with well-rounded tones (e.g. "How are you?"). Speak in front of a candle in your usual voice and notice how the candle flickers with your breath. Position the candle at the limit of when you make the candle flicker in a regular voice. Next try the breathing control and repeat the phrase. Do you blow the candle out? You should be able to keep moving the candle back a little more until you can affect it without shouting. Now try the breathing control again, but practice retaining your breath for longer periods, eventually of up to 40 seconds or more. When you can maintain a long breath, begin articulating a paragraph of text and imagine projecting to the back of an auditorium.

Mannerisms and Posture

Posture and mannerisms project a message about how we feel. Memorable public speakers try to project confidence and credibility. This can be practiced and learned. In the following chapter we discuss non-verbal communication in more depth, but a synopsis will be covered here as it relates to public speaking.

First know how you look in front of an audience. Have someone videotape you while you present. This should show you the aspects of your "body language" that may be sending the wrong message. Standing straight without slouching are all part of a confident posture. If you suffer from not knowing what to do with your hands, hold a pointer or put one hand in your pocket (have your pocket empty so you don't play with coins or keys) or behind your back. Find what works for you. Try to avoid both hands in your pockets or having your arms folded in front of you. This tends to project nervousness or even hostility.

If you can and it feels comfortable, don't be afraid to walk slowly about the front of the audience. Again, go with what works for you. Getting away from a single spot on the floor can be a useful stress-buster. If you are using a podium or lectern, try not to grip it tightly or hide behind it. Practicing before friends will give you feedback to develop your confidence. Remember that the aspects given above go to making a good presenter. The more you do it, the easier it will become. Box 1 emphasizes some of the characteristics of various types of speakers. Aim to be credible and comfortable.

OVERCOMING ANXIETY ABOUT PUBLIC SPEAKING

Most people can identify with the fear of presenting before an audience. Much of this fear is about something going wrong and embarrassing you. In a survey of the American public, speaking before a group was the highest ranked (41% of total) pervasive fear (Wallechinsky et al. 1977). When giving a presentation we can become so fearful of things going wrong that we freeze and make it a self-fulfilling prophesy. What is the worst that could possibly happen? You forget everything you wanted to say, your notes are blown away by a sudden howling wind, all of your visual aids suddenly evaporate, and all your clothes suddenly fall off revealing your nakedness in public! So, perhaps you can't quite remember what you wanted to say—stop worrying about it and say what you can. After all, if you were well-prepared you can always refer to your outline or use your visual aids. The rest is just unfounded fear. And if the worst should actually happen, the audience will be sympathetic and understanding.

Stress and fear are a normal part of our lives and should not be construed as negative aspects of who we are. But, fear can be either rational or irrational. Rational fears prevent us from stepping off cliffs and getting into a cage with a lion. Irrational fears, however, tend to be constructs that we have built into our individual realities because of past experiences. It is necessary to recognize our fears and identify them before beginning to resolve them. Box 2 contains a self-test to evaluate your fear with respect to public speaking.

Whatever the source of our fears, the body reacts with a basic flight or fight mechanism and floods our systems with adrenaline. It is this hormone that makes us feel jittery and anxious, and gears the body to action. Unfortunately, with irrational fears, we tend to have no outlets and it is this non-action that creates the poor performance typical of bad presentations. Understanding our fears thus allows us to refocus all that energy and actually use it to good effect rather than suppressing it and impairing ourselves.

The first thing to overcome one's fear of public speaking is to understand apprehension and to prepare for it. Fear makes your body produce adrenaline, so that you are prepared for "fight or flight." This is useful when you are about to face a tiger, but it is not helpful when facing an audience (of any size) to give a presentation. You are not supposed to fight with the audience, nor are you really allowed to run away. You have to do something else with the energy that is being mobilized by your body. The following section will give you some ideas on how this can be best achieved for you. No single technique is right for everyone. Only by knowing yourself and understanding why you are stressful can you develop your "own style" that lets you become confident and relaxed.

Five different fears have been identified with public speaking. Understanding these fears, and the processes for overcoming them, is important to overcoming speaking anxiety (Beatty 1988).

1. **Novelty**—Be prepared mentally for new situations. Not being prepared for novelty, such as doing an electronic presentation when one is unfamiliar with computers, will make you anxious.

2. **Subordinate status**—A feeling that you are inferior to the audience. Think positively about yourself and understand that the audience is there to listen to you because they want to hear what you have to say.

3. **Conspicuousness**—Feeling unprepared to be the center of attention. In a conversation with friends you would not mind being the center of attention. Visualize your audience as just a larger group of friends who want to listen to what you have to say.

4. **Dissimilarity**—A feeling of having nothing in common with the audience. Understand and analyze your audience. They are there to listen to you. If you have targeted your audience correctly, you will be giving a talk that they want to hear.

5. **History**—The personal knowledge that you have been anxious before. Think of all your previous experiences as learning sessions that are preparing you to be better. After all, none of us were able to walk as a baby without lots of practice. The more you talk in public, the more your confidence will grow.

Finally, relaxation is important to giving polished presentations. Box 3 lists several powerful ways to relax the body and mind.

Box 1: Types of Speakers

What kind of speaker do you want to be? Some common types of speakers are identified below; undoubtedly you have encountered them. In the future, when you are attending a presentation, observe the speaker and categorize him or her. What can you learn about your own speaking style?

The Shrinking Violet

This kind of speaker does not address the audience; the only thing that can clearly hear the presentation is the speaker's shoes since that is where the speaker spends most of the time looking. This type of speaker has lost the audience after the first few lines. If the audience does not feel involved in the speech or is bored, they will not listen. The same is true for a speaker that uses the words "Um, erhhh, urhm" frequently—it gets too hard to listen and concentrate.

The Zealous Animator

This kind of person addresses walls, ceiling and assorted inanimate objects in the room, and they do so with great theatrical flair by gesticulating wildly. This has the same effect as the shrinking violet in losing the audience, for now the audience is more interested in the arm-waving antics of the speaker rather than the content of the talk.

The Fire & Brimstone Preacher

A high-energy approach using high speed delivery and high-volume is used to present the topic. This person may also pound the podium or nearby table which leaves audiences shell-shocked. The speaker may also talk "at" the audience in an overbearing authoritative way. While this may have some effect with evangelical-type speakers in the correct setting, it tends to alienate most people.

The Credible Speaker

A relaxed, assured, confident presentation by a person who seems to enjoy public speaking. This type of speaker appears to enjoy themselves while presenting and seems to be talking "with" the audience in a conversational style rather than just "lecturing" to the audience.

Box 2: A Speaker Apprehension Test

The following "apprehensiveness self-test" (McCroskey, 1982) will help in understanding what are your strong and weak points in dealing with public speaking apprehension.

Directions: The questionnaire has 24 statements on how you feel about communicating with other people. For each statement assign yourself the appropriate number of points that best corresponds with how you feel. There are no right or wrong answers, only what you feel. This will give you the best understanding of yourself.

1=Strongly agree, 2=Agree, 3=Undecided or don't know, 4=Disagree, 5= Strongly disagree.

Many statements seem similar but measure different aspects. Work quickly and record your first impression.

1. I dislike participating in group discussions.
2. Generally, I am comfortable while participating in group discussions.
3. I am tense and nervous while participating in group discussions.
4. I like to get involved in group discussions.
5. Engaging in a group discussion with new people makes me tense and nervous.
6. I am calm and relaxed while participating in group discussions.
7. Generally, I am nervous when I have to participate in a meeting.
8. Usually, I am calm and relaxed while participating in meetings.
9. I am very calm and relaxed when I am called upon to express an opinion at a meeting.
10. I am afraid to express myself at meetings.
11. Communicating at meetings usually makes me uncomfortable.
12. I am very relaxed while answering questions at a meeting.
13. While participating in a conversation with a new acquaintance, I feel very nervous.
14. I have no fear of speaking up in conversations.
15. Ordinarily, I am very tense and nervous in conversations.
16. Ordinarily, I am very calm and relaxed in conversations.
17. While conversing with a new acquaintance, I feel very relaxed.
18. I am afraid to speak up in conversations.
19. I have no fear of giving a speech.
20. Certain parts of my body feel very tense and rigid while I am giving a speech.
21. I feel very relaxed while giving a speech.
22. My thoughts become confused and jumbled when I am giving a speech.
23. I face with confidence the prospect of giving a speech.
24. While giving a speech, I get so nervous that I forget facts I really know.

Scoring: Follow the instructions. This questionnaire has one total score and four sub-scores as given below:

Subscores:

Group Discussions:
Begin with 18 points
   then add scores for items 2, 4, and 6
   And subtract scores for items 1, 3, and 5

Meetings:
Begin with 18 points
   then add scores for items 8, 9 and 12
   And subtract scores for items 7, 10, and 11

Interpersonal Conversations:
Begin with 18 points
   then add scores for items 14, 16, and 17
   And subtract scores for items 13, 15, and 18

Public Speaking:
Begin with 18 points
   then add scores for items 19, 21, and 23
   And subtract scores for items 20, 22, and 24

Each subscore ranges from 6–30. The higher the score, the greater the apprehension. Any score above 18 in any category indicates some degree of apprehension. To obtain your total score, add the four subscores together. A total score above 70 may indicate a high degree of shyness which can be overcome by practice and positive thinking. Joining a speakers group that can give positive and constructive feedback can help overcome many of the fears of public speaking.

Box 3: Relaxation Techniques

Many times our stress can be actively reduced or refocused. Stress can remain positive when we take time to relax after or even during a challenge. The techniques that follow may require some practice, but if made into everyday habits, can be effective in reducing stress in your life.

Deep Breathing

A.   This exercise is central to all relaxation methods. Inhale slowly (5 seconds) through the nose and expand the lungs completely. Exhale slowly (5 seconds) through the mouth. Repeat several times.

B.   Repeat A, but this time put your hands on your stomach and as you inhale, expand your stomach and your lungs. Your hands should rise if you are doing it correctly. A good addition is to think or say the word "calm" being slowly exhale with your breath. Both of these methods can be done anywhere at anytime.

Mind Clearing and Imagery

A.   This practice is also basic to many relaxation methods. Find a quiet place without distractions. Get in a comfortable position. Close your eyes and do deep breathing. Form a mental picture of a peaceful place or event. Concentrate on that image while you breath in a relaxed fashion. Stretch upon completion of the exercise.

B.   Repeat A, but continue the image and allow the mind to run free. Use your imagination to take a mental vacation of your choice whenever you need to relax.

C.   Day dreaming is a healthy and useful technique if done so it does not interfere with the task at hand. It should be done at a time set aside to relax.

Stretching

Stretching various parts of the body can cause those muscles to relax. Concentrate on the muscle groups that bother you most. Relax, breath deeply, and slowly stretch the muscles that are knotted or tight.

Moving

Many times we will stand behind a podium, thereby creating a barrier between the speaker and the audience. A good speaker can still make the audience feel involved in the presentation and also control any nervousness by channeling it into proper projection and inflection. A nervous speaker, however, will make the podium a haven to hide behind. Nervous speakers also will tend to exhibit other features such as involuntarily shaking the podium while gripping the sides with white knuckles, fidgeting and jerky mannerisms, and shuffling from foot to foot.

In many cases, it is extremely useful if the nervous speaker can channel some of that nervous energy in another direction that gives them some control. One such technique, if space is not limited, is to walk around slowly and deliberately, with varied stops. It can also help in eye contact since as the speaker turns, his body will automatically scan across the audience. Although not recommended in all cases, it does afford the nervous speaker a way to channel energy into his feet and away from the rest of the body.

Positive Mental Attitude (PMA)

A positive and healthy approach to life is the best way to insure a continuation of the same, as well as reducing the chances of developing a stress-related illness. Positive thinking is essential. Tell yourself "I CAN" rather than defeating yourself with negatives before you begin. Do not say "I can't." Remember, BE POSITIVE. Communicate and share your experiences with friends, family or members of a support group. "Getting it out" reduces stress. If a negative stress occurs in your life, there are three basic options to deal with it. First, try to change the system that is causing the problem so it no longer exists. Second, develop new attitudes or behaviors to cope with the problem in a more positive way. The third option is to quit what you were doing in order to eliminate that stress in your life. Be an optimist rather than a pessimist. See that glass as half full rather than half empty.

REFERENCES AND FURTHER READING

Arredondo, L. (1990). *How to present like a pro!: Getting people to see things your way*. McGraw-Hill.

Beatty, M. (1988). Situational and predispositional correlates of public speaking anxiety. *Communication Education, 37,* 28–39.

Bender, P. U. (1995). *Secrets of power presentations*. Firefly Books.

Berkley, S. (1999). *Speak to influence: How to unlock the hidden power of your voice*. Campbell Hall Press.

Booher, D. (1994). *Communicate with confidence: How to say it right the first time and every time*. McGraw-Hill.

Condrill, J., & Bough, B. (1999). *101 ways to improve your communication skills instantly*. Goalminds.

Decker, B., & Crisp, M. J. (Eds.). (1997). *The art of communicating: Achieving interpersonal impact in business* (Revised ed.). Crisp Publications.

Fetzer, S. (1984). *Improving your speaking voice. Winning with words*, pp. 79–107. World Book Encyclopedia, Inc.

Krasne, M. T. (1997). *Say it with confidence: Overcoming the mental blocks that keep you from making great presentations & speeches*. Warner Books.

McCroskey, J. C. (1982). *Introduction to rhetorical communication* (4th ed.). Prentice Hall.

Mckerrow, R. E., Gronbeck, B. E., Ehninger, D., & Monroe, A. H. (2000). *Principles and types of speech* (14th ed.). Longman (An Imprint of Addison Wesley Longman).

Rozakis, L. E. (1999). *The complete idiot's guide to public speaking*. McMillan Distributers.

Swets, P. W. (1992). *The art of talking so that people will listen: Getting through to family, friends, and business associates*. Prentice Hall.

Timm, P. R. (1997). *How to make winning presentations: 30 action tips for getting your ideas across with clarity and impact*. Career Press.

Urech, E. (1998). *Speaking globally: Effective presentations across international and cultural boundaries*. Kogan Page Ltd.

Wallechinsky, D., Wallace, I., & Wallace, A. (1977). *The people's almanac presents the book of lists*. William Morrow, p. 469.

Wilder, L. (1999). *7 steps to fearless speaking*. Wiley & Sons.

Woodall, M. K. (1996). *Thinking on your feet: How to communicate under pressure*. Professional Business Communications.

Zarefsky, D. (1999). *Public speaking: Strategies for success* (2nd ed.). Allyn & Bacon.

14

COMMUNICATING WITHOUT WORDS

When discussing communication, we intuitively think of the language of words. Words are powerful conveyers of meaning and they do matter in getting your message across. But, there is a whole other realm of communication, one that transfers meaning between sender and receiver without using words. This is the realm of non-verbal communication (also referred to as "subtext").

It has been estimated that between 60% and 95% of the meaning transfer in a communication system is accomplished through non-verbals. This is more than "body language." Non-verbal communication includes physical and psychological signals, subtle inferred meanings in the motions of sender and receiver, implications of unspoken words, and buried cultural expectations present when communicating. Indeed, we often rely more on non-verbal cues than the actual words used. For example, a friend says to you "I studied all night" but stresses the word "all," laughs and rolls her eyes. You immediately know she is being sarcastic, and in fact did not spend much time studying.

It is not just what you say with words, but how you say them. Non-verbal communication is all the ways we add or change meanings of the words actually spoken.

In this chapter, we discuss some of the ways we are socialized in the use of non-verbal communication. We hope to make you aware of many non-verbal forms of communicating to help you understand your own patterns and those of people with whom you interact. Knowing non-verbal language, however, is not a reason to practice deception. It should be used to enhance your spoken message to improve clarity and increase credibility. It must be emphasized that non-verbal language is dependent on culture, since many symbols and movements can vary depending on the user or receiver's experience and semi-conscious intent. All cultures exhibit non-verbal communication and often in a similar fashion.

KINESICS: PHYSICAL MOVEMENT

Whenever a person feels confused or senses a contradiction in a conversational setting, it is often because the non-verbals, from certain physical movements and body actions, are not matching the spoken words. For the purposes of this chapter, kinesics (the study of gestures, body language, and facial expressions, especially in combination with speech) can be separated into three aspects, though all three will probably occur at the same time. You are often aware of this at a subconscious level, and so are often reacting at an emotional level to the kinesic messages you are receiving.

- **Body language**: It is estimated that we use body language two-thirds of the time we are in personal communication, usually subconsciously. The dynamics of body language, a sequence of motions and gestures, lend meaning to the subtext. Examples of body language are leaning towards someone you find interesting, turning your shoulder and back towards a third person you are "excluding" from a three-party conversation, and crossing your arms in front of you to create a barrier. Becoming aware of the movements can help you to communicate clearer. By matching your body language to your spoken language you will increase your credibility.

- **Gestures**: Gestures are dependent upon the context and situation in which they occur. The same gesture may have several meanings at different times and places. Handshakes are the basic measure of respectful greeting in the United States. How you shake can affect how the other person feels about you. In Japan, the angle of the bow determines whether you will insult the other person or demean yourself, but this is usually waived for non-Japanese, who need only bow to show respect. A single wave of the back of a hand may be seen as an insulting dismissal or just a "no problem" comment. Raised eyebrows can be a sign of surprise especially when accompanied with a smile, but one raised eyebrow with a serious expression can be a sign of disapproval or disagreement. If it is a boss-to-subordinate situation then it can become a power play, or at least diminish the confidence of the person speaking. Naturally, most cultures have some gestures that have a derogatory nature about them and should not be used in any professional circumstances, even when angry.

- **Facial Expressions** (including eye movement): Raising eyebrows or smiles can usually be friendly but can mean disrespect or aggression if held too long. In the American culture, whenever someone holds eye contact for more than about one second something must happen—either a smile, or a salutation of some form is customary.

PROXEMICS: PERSONAL SPACE

We all have an individual zone of personal space that surrounds us. These zones are where we position ourselves in relationship to others. Therefore, how we situate ourselves is critical to our personal comfort, yet we need to be aware of how we impact other peoples' zones. In reality this bubble seems to be larger in the areas that we can see, especially extended to the front and somewhat at the sides. Examples where proxemics can be readily seen in action are in the elevator, how people arrange themselves on chairs in a waiting room, or where people sit at a bar.

Take some time and watch how people arrange themselves when in groups in various locations and at different functions. The distances given below are based on research by Edward T. Hall (1963).

- **Intimate Zone**: 0-18 inches

  As the name suggests, this zone is reserved for highly personal relationships. Being uninvited within the intimate zone can create a feeling of anxiety in the other person and lead to aggressive reactions. Think about a crowded elevator. Did you feel anxiety about people standing close to you? It will become easy to spot those people who are comfortable with each other, those that need their distance, or are strangers.

- **Personal Zone**: 18 inches-4 feet

  Most people in a professional situation will adopt the personal zone and maintain the distance. Try talking to someone and then slowly step back and to the side. The other person should follow your pattern in order to maintain the same distance. To talk from too far away is to appear too cold and distant, and too close is to be invasive. We naturally feel our correct distance even though it is culturally learned. If your comfort distance is closer than someone else's, you will notice signs of discomfort in them. That is your cue to slightly increase the distance between yourselves.

- **Social Zone**: 4-12 feet

  In a group, people will tend to space themselves equally several feet from the speaker, yet close-up within the group. Since individuals in groups are usually standing side-to-side they do not require as much space as they would face-to-face. If the group is larger and feeling "crushed" there will be much unconscious negotiation of space between each of the members of the group. The less comfortable members of the group may stand far to the back to increase their space or even leave the group. If this happens then the speaker might want to actively get the group into the public zone where each member can claim whatever space they need.

- **Public Zone**: 12 feet and more

  In this zone, the distance of the speaker is rarely a problem except that they are now farther from the audience and thus are in danger of losing that special connection that allows personality. The speaker needs to become more dynamic to retain a connection with the audience. In the audience however, as the group gets bigger, there is increased negotiation on space. Seating arrangements in a theater or a hall need to be carefully thought out to ensure that everyone can have adequate space.

Be aware of the distance that people need to be comfortable. If they are feeling anxious they are probably not listening well.

SEMIOTICS: THE SCIENCE OF SYMBOLS

Semiotics can be thought of as the application of non-verbal communication to influence an audience. Here, it is applied kinesics and proxemics, the bringing together of non-verbals to create a favorable impression on a listener. Semiotics could also be used to create a pleasant environment for an important meeting. Below are several ways of using semiotics.

- **Physical appearance**—You have probably heard that "neatness counts" and "dress for success." Appearance varies for different ethnic groups/cultures. While it may not be "correct" from a multicultural perspective to judge a person by their appearance, it would be naive to not believe that credibility, professionalism, and even respect are gauged by the type of clothes that one may wear in any specific setting.

- **Smell/Taste**—These are probably the most overlooked nonverbal factors, yet they can have significant effects on moods and psychological associations. Perfumes and fragrances for example, while pleasing in some situations like a romantic evening or meditational setting, can become quite distracting in a professional setting. Certain tastes and smells can evoke intense memories. In general, it is advised to be careful when setting up a situation involving odors or tastes.

- **Eye contact**—The way we look at people and how we use our eyes during personal interactions indicates a lot about what we are thinking about. In Euro-American cultures, it is usually expected to hold eye contact while talking to someone because this indicates that you are listening. However, eye contact between strangers becomes uncomfortable if held for more than a second or two. Some cultures use continued eye contact as a form of aggression (an invitation to fight) while others use it as a sign of warmth and respect. If you are involved with different cultures, it behooves you to learn the cultural nuances for your own benefit. Begin by noting how you use eye contact in your everyday communication.

- **Smiles**—Smiling is believed to be universal in cultures all over the world (Konner 1987). This is one facial expression that has the capacity to warm up any situation and enhance any presentation. It usually conveys warmth and positiveness. Of course, like any action, it can have a negative connotation as with "the crocodile smile" where the person smiling is insincere. Usually a whole nonverbal review of the smiling person should reveal whether the smile is genuine or not.

- **Haptics** (physical touching)—If in doubt, keep your hands to yourself. If you use haptics in your everyday body language, become aware of how you use them and in what situations so that you can use them to enhance your personal interactions for positive results. If you do not use haptics be careful of beginning them until you understand the full implications of what a touch means or how it can be misconstrued. In today's society, many haptics may be misconstrued by people as harassing or as plays for dominance. By all means do use, or learn to use, haptics for they are one of the most influential non-verbal factors that allow us to show caring and empathy. The frequency of touching is also a positive and strong aspect of haptics, if done suitably for a given situation.

- **Gestures of respect**—These may be greeting gestures such as correct hand shakes or bows. They also may be simple things like type of clothes. Wearing the right clothing for a specific situation (e.g. job interview) signals respect for the other person.

- **Note taking**—compliments the speaker because you appear to be interested in what they say. Be sure it is acceptable in the culture and under the situation.

- **Dominance factors**—Be aware of aggressive power play. You should have realized by now that how you use non-verbals in a semiotic way can be either negative or positive. If you feel negativeness is occurring, try to assess your non-verbals or those of the people with whom you are communicating to decide if one of you is trying to be dominant over the other. If you are having a positive interaction, assess the non-verbals again to understand what you are both doing that makes it so.

- **Setting the scene**—Create the environment for success with the correct setting. Imagine trying to have a romantic dinner on a busy railway station platform! Obviously, a more remote and subdued lighting environment would be more "aesthetic" and appropriate. Similarly, think of having a business luncheon in a pleasant restaurant. It usually confers a less formal atmosphere and more open communication.

- **Seating at a conference table**—Who sits at the head of the table and who sits next to them? How is the power controlled just by the seating arrangements? Are you trying to enforce your dominance or do you wish to create more equitability? If you create a seating arrangement of "us" opposite "them," then the table becomes a barrier. Preferential seating can also create enmity within the group, or alternatively establish the "pecking order" and reduce potential conflict. As usual, knowing your audience is essential.

PARALANGUAGE

There are several ways that we use spoken language to give a different or additional message than the one we are actually saying. These are:

- **Articulation**—speaking clearly and confidently increases credibility
- **Pronunciation**—helps the audience clearly understand what has been said, thus improving communication
- **Emphasis**—putting the emphasis on certain words gives the words a different meaning and brings the audience's attention to those words. Ways that this often happens are:

  Pitch—Using a higher pitch for positiveness or a low pitch to express negativeness.

  Rate—speaking slowly to bring more emphasis to every word, or fast to gloss over the words.

  Timbre—imposing a harshness or softness in the words, or whispering instead of shouting.

  Pauses—this can give dramatic effect to the words just spoken.

Psycholinguistics

This tends to be the use of all the above to create unspoken meaning or even dominance factors. Think of the following sentence: "Jim has done a great job on this assignment." Say this sentence three different ways, but use non-verbal communication and varied voice inflection to give a different meaning each time to the sentence. Think how voice inflection and non-verbal communication can alter the meaning of a message. If there is a conflict between the implied message and the verbal message, the implied (non-verbal) message is more likely to be believed.

Presupposition Statements

Elgin (1980) emphasized how the spoken word can be changed in many ways to create this hidden meaning. Read through the list and then think of the hidden presupposition that is represented. How often have you been a receiver of a negative presupposition that has left you angry and unsure of why you are so upset? Alternatively, do you use presuppositions as a form of your own and wonder why people are so easily upset by you? Some examples, derived from Elgin, are given below. The word(s) to accent is bolded in each line.

- If you **really** cared, you wouldn't go. . .
  Presupposition: You don't care!

- If you **really** cared you wouldn't **want** to go. . .
  Presuppositions:
  > You don't care.
  > You have the power to control your decision.

- Don't you even **care** about your weight?
  Presuppositions:
  > You don't care.
  > You should care, it's wrong not to.
  > You should feel guilty and rotten.

- Even a **man** should be able to understand **biology**.
  Presuppositions:
  > There's something wrong with being a man.
  > It doesn't take much to understand biology.
  > You should feel stupid and guilty.

- A person who **really** wanted to save money would spend it **carefully**.
  Presuppositions:
  > You don't want to save money.
  > You spend money carelessly.

Elgin recommends that in order to defend yourself successfully against negative presuppositions, you need to identify the presuppositions and then be sure to address them. In the last statement, you would need to address **both** your desire to want to save **and** your wish to be a careful spender! To ignore either means you have already lost the argument!

METAPHORS

Some cultural stories may hold a hidden meaning, or implied nature of something, that is only known because of being a member of a specific culture. An example of this is the Navajo coders or "code talkers" in the Pacific theater of World War II. These coders spoke to each other in the Navajo language over open radio channels, yet the only people who could understand the message were other coders who knew the tribal myths and stories and could identify what was implied in the messages. The enemies were never able to crack this code. In any culture, there is a great degree of language that uses metaphors to set up a concept for the receiver (Lakoff & Johnson 1980). If using metaphors with a specific audience, ensure that they understand the metaphor's real meanings and not read too little or too much into the metaphor. By virtue of metaphors being analogies, they are not exact and it is the "extra" parts of the metaphor that may create problems.

As an example, a natural resources professional may make reference to a situation being a "tragedy of the commons." Audience members who are unfamiliar with this work by Garrett Hardin will not understand the point, however. So while metaphors are useful to create analogies, they need to be based on shared understanding.

CULTURAL IMPLICATIONS

Non-verbal communication is dependent on the cultural context in which it is delivered. This includes different ethnic and national cultures. Cultural considerations were also covered in Chapter 12. For now, some simple rules for using non-verbals within cultures other than your own are:

- Keep body language simple to be less confusing.
- Know your audience when giving an oral presentation. Be sure your non-verbals and language mean the same thing to the audience that they do to you.
- Understand that non-verbals are dynamic (a sequence of movements) and you should be wary of trying to interpret singular actions. Individuals may have developed unusual personal meanings for some actions, yet will still comply to cultural norms for overall body language.
- Develop a conscious sense of the unspoken meanings around the words in a conversation. This is an ideal way to begin to reduce conflict and dominance factors, and thus avoid inadvertently offending or hurting people.

CONCLUSION

Nonverbal communication is a critical yet often overlooked aspect of delivering a message. Consider the difficulty of holding a meeting between parties with little trust—perhaps between a community action group and a large corporation. What actions could be taken, based on what you have learned in this chapter to create an atmosphere of trust?

REFERENCES AND FURTHER READING

Breasure, J. (1982). *Non verbal communication skills.* Advanced Development Systems.

Dimitrius, J., & Mazzarella, M. (1999). *Reading people: How to understand people and predict their behavior, anytime, anyplace.* Ballantine Books.

Elgin, S. H. (1980). *The gentle art of verbal self defense.* Prentice Hall.

Fast, J. (1970). *Body language.* Pocket Books.

Fast, J. (1994). *Body language in the workplace.* Penguin Books.

Hall, E. T. (1963). Proxemics: A study of man's spatial relationship. *In man's image in medicine and anthropology.* International Universities Press.

Konner, M. (1987). The enigmatic smile. *Psychology Today,* 21 (March, 1987), 42–46.

Lakoff, G., & Johnson, M. (1980). *Metaphors we live by.* The University of Chicago Press.

Nierenberg, G. I. (1982). *How to read a person like a book* (Reissue edition). Pocket Books.

15

USING VISUAL AIDS

This chapter deals with visual aids used to enhance oral presentations. It is intended to guide planning and presentation of visual aids; actual construction of visual aids is covered in other texts, some of which are recommended at the end of the chapter. What sets off a visual aid from a visual message is how is it applied and used. Visual aids, as they are explained here, are meant to enhance a personal presentation, while a visual message is meant to present a stand-alone message in the absence of a presenter (e.g., a park interpretation sign or an advertisement).

The old adage is that a picture is worth a thousand words, but it is important to understand that this valued picture be the right picture, presented well, at the right time, or else it will detract from the presentation rather than support it. A visual aid which is well-designed and suitably placed can give much more clarity to a message in a shorter time than a lengthy set of words. Combining both words and visual aids greatly increases the impact. Visual aids are important because of differences in the way people perceive messages. Some audience members will react best to verbal information, some to visual, and some to written material. Thus presenting messages in multiple formats broadens the audience impact. People tend to remember more of what they see and hear. So visual aids supporting verbal information act to dramatically increase the impact of a presentation. Furthermore, visual aids increase an audience's attention span, and can, when used properly, add structure to information to facilitate understanding.

There are many types of visual aids. Using the right one in a particular situation relies on an understanding of the strengths and limitations of each. In essence, anything that visually depicts a message either directly, subtlely, or semiotically is a visual aid. As a departure point, the presenter should draft up the presentation first and then find visual aids that fit and enhance the presentation and not vice-versa. Remember, the message is what is important. The visual aids are just enhancements to help the audience get the message.

A key point when using visual aids can be summed up in the term "**Say dog, see dog**." If you are talking about a dog, then the visual aid being used should be a dog. Likewise, if the visual aid you are displaying is a dog, then you should be talking about a dog. This rule is often neglected in presentations. For example, if you get a puzzled look from your listeners when you are showing a picture of a bison, and you are talking about the Black Hills of South Dakota, then be aware that the audience may not be making the connection that you may be taking for granted. Yes, there are bison in the Black Hills, but the audience may not know this and they may be trying to understand why they are not seeing a picture of the Black Hills. At this point the audience's attention is waning and you will not be as effective in your presentation.

Visual aids should be used during a presentation to achieve specific objectives. Here's how they add effectiveness and professionalism to your presentations:

- Use an eye-grabbing title with an image to get the audience's attention. The opening needs to break the audience's preoccupation and draw them into the talk. The opening visual aid can act as "feedforward."

- Select visual aids that guide the group's thinking to pre-determined conclusions. It is essential to know the audience so that they do not misinterpret the meaning of the visual aids.

- Emphasize key points, not every sentence.

  - The most common problem with visual aids is that people tend to over-complicate their presentations. Don't put the whole story on the screen—just key words or simple graphics—enough to help people keep track of the idea.

  - Give the audience a reason to listen to you. If a visual aid is complete enough to use as a handout, then use it as a handout.

- Present complex, detailed data in understandable ways.

- Explain new concepts, pictures and diagrams to help clarify the details.

Visual aids should be graphic. People think in pictures. If using word-visual aids to reinforce the message, keep the words to a minimum. The visual aids should enhance the presentation, not distract from it. If you have a powerful and flashy visual, give the audience several seconds to absorb it and then continue once the "awe" factor has worn off. Color is useful and can enhance visual aids, but try not to be distracting. The colors should draw the eye to key points that the speaker is addressing. With today's superb and easy-to-use computer graphics it is simple to get carried away with too much color.

Other essential things to think about with your visual aids are:

- Is it clear? Is it obvious at a glance what the visual aid is trying to communicate?

- Is it readable? The arm's length rule—the image should be as readable at a distance as if it is at arm's length. Hence the distance of the farthest part of the audience from the visual aid needs to be anticipated so that the graphics or fonts are large enough to see as though they were only an arm's length from them.

- Does it communicate a single idea? Is everyone able to focus on that idea? If there are too many points on a visual aid, then the audience may be looking at the wrong point while you are speaking.
- Is it relevant? Does it make a point that fits in with the presentation? Don't reveal it until you're ready and remove it when you're finished.
- Is it interesting? Does it keep the audience's attention? Graphics should capture attention.
- Is it simple? Is it too cluttered with pictures, graphs, colors, borders and lines? Make the focus be the point at which you wish the audience to look.
- Does it support the content? Remember "Say dog, see dog."

HANDOUTS

Handouts are visual aids that you distribute to the audience so that they all have their own copies. You may choose to distribute an outline of your presentation so that the audience can follow your discussion. Or you may want to give them essential information so that they do not have to take notes while you are speaking. You may also give the audience a summary of your discussion, especially if there will be several speakers and you want the audience to have a reminder of your talk. Finally, some speakers use handouts to give the audience supplementary material that can be used for later reference.

When deciding when to distribute handouts, you do not want to disrupt your presentation or distract the audience. The following are the preferred times to distribute handouts:

	Prior	During	After
1. As an outline	Best	Poor	OK
2. Material essential to discussion	Best	OK	Poor
3. As a summary	Poor	Poor	Best
4. Supplementary material	OK	Poor	Best

Be careful that you do not overdo the volume of handouts you give to the audience. This will tend to confuse your audience and distract from your presentation. If you have many sheets, bind them into a booklet and then use the booklet as a workbook that complements your presentation. Leave white space on each page so that audience members can enter their own thoughts about your presentation. Color code sections where appropriate. This is especially true when the handouts cover complex diagrams, or there are many pages covering different types of topics.

OVERHEAD PROJECTOR AND TRANSPARENCIES

This is still the most widely used method of presenting visual aid information because of its ease of preparation and versatility. Overheads can be easily prepared with a computer and printed on any printer or by duplicating onto a transparency sheet using a photocopier. Be sure you have the right transparency sheets for the appropriate device. For instance, laser printers and photocopiers use transparencies able to take high temperatures, while inkjet printers use transparencies with a special surface able to receive the ink.

There are advantages and disadvantages to transparencies:

Advantages

1. You can face the audience—this should be true of all types of presentations.
2. Normal room lighting allows the audience to look easily at the presenter and the overheads.
3. Flexibility in preparing, editing and revising since they are easy to make.
4. Since the overheads are always close at hand you can answer unanticipated questions which may refer to an overhead from the presentation.
5. They are easily portable from one place to another, usually not taking up more than a single file folder in size and weight.

Disadvantages

1. Burned-out bulbs can be a nuisance. Many projectors have spare bulbs in them, but you need to know how to change the bulb without electrocuting or burning yourself.
2. Glare from some overhead lights may cause the overhead to wash out. You can often just switch off the offending light or reposition the projector better.

Techniques for using transparencies and an overhead projector are discussed in Box 1.

SLIDES

Slides are an economical and, depending on the resources at hand, excellent way to produce quality graphics and visual aids in a short time. Resources that are needed are a high quality 35mm camera, slide film, a place to develop the film and a slide projector with carousel. A close focus micro lens is essential if good copy work is desired. Many professional camera shops can develop and mount the "positives" in about an hour, although black and white film can take substantially longer. Slides are especially good for large groups. Many computer programs now available can also produce slides of photographic images and computer generated graphics.

Slides can create an effective presentation. But, you'll find a healthy dose of practice is required to present a memorable slide show. The following ideas can be used to create professional quality slide presentations:

- Prepare the audience before the room lights are dimmed or shut-off. Feed-forward on your topic is particularly important here to get the audience thinking about what they are to be seeing.

- Leave some lights on in the room if possible or use light dimmers if available. This will help the audience take notes if appropriate, or at least stop them from dozing off too readily in the anonymous darkness. It is important for the presenter to speak quite dynamically since the usually non-verbal enhancers are subdued in the darker room.

- One of the biggest problems when using slides is using a room where no shades or curtains are available to shut out light. The light glare can wash out the slides to the point of making them useless. Visit the room ahead of your presentation to be certain slides can be used.

- Use an average of one slide per 15–20 seconds. If you are talking about one slide for more than 40 seconds then think about increasing the number of slides on the points being covered. If using a sequence of slides as illustrations (e.g. monkeys to be found in the primate enclosure of a zoo), allow several seconds per slide so that the audience can absorb the visual. Pointing out one or at most two keys points per slide in the sequence will focus the audience on what is essential in the slide.
- Break presentation into segments of about 5–9 slides that cover a particular topic. Too few slides and the audience may start to feel confused; too many and they may start to get bored.
- Review the key points of the presentation at the end.
- If there are large amounts of information being presented, then prepare a worksheet for the audience to review the presentation and to summarize their own thoughts about the information.

Additional techniques for using slides are given in Box 2.

FLIPCHARTS, CHALKBOARDS AND WHITEBOARDS

These forms of visual aids are especially good for interactive sessions where the presenter has to write information or develop lists generated during discussions/interactions with the audience. One important point about write-on media is to avoid talking completely to the board or chart when writing. While writing you should stop and look at the audience to reinforce that you are talking to them. If you have so much to write that you have too much "dead time" when you are not speaking, then use prepared media or use an assistant to do the writing. You will want to carry your own supplies of pens and chalk.

Flipcharts

The best uses for flipcharts are: 1) making lists; 2) outlining steps in a process; 3) sequencing ideas; 4) drawing simple sketches; and 5) recording group work such as brainstorming. Advantages and disadvantages to using flip charts are:

Advantages

- Spontaneous with little preparation needed.
- Easy and inexpensive to use—requires only a stand, flipchart paper and pens.
- Can be used in normal room lighting.
- Used sheets can be easily posted around the room for re-emphasis during ongoing discussions.
- The pages can be saved as a record of the discussion.

Disadvantages

- Turning your back to audience to write is necessary.
- Not good for detailed information.
- Time-consuming.

- Difficult to carry.

Tips for Good Flipcharts

- Use two flipcharts—one for prepared materials; the other for spontaneous comments and lists. Or use one for questions, and one for responses.
- Keep statements short and simple to minimize writing.
- Write a title on every page—it adds impact and helps organize.
- Colors help organize the information. The colors with the greatest visibility are black, blue, and green, in that order. Avoid purple, brown, pink, and especially yellow. Red should only be used as an accent color—for bullets, underlines, arrows, etc. Key words may be written in red when everything else is blue or black.
- Two colors in combination on a chart are better than one. Three are OK if done carefully and with purpose. More than three makes it difficult for the audience to pick up points of emphasis. Combinations to avoid because of contrast and visibility are red and green, orange and blue, or yellow with any other color except black. Good contrasts are red with black or blue, and green, yellow or blue with black (and vice versa).
- Lettering needs to be consistent and neat in style when used on a chart, otherwise it can get distracting. Use print, not script. Use upper and lower case. Using grid lined paper will help keep lettering straight and allow you to line up margins, subheadings, and bullets.
- Leave generous amounts of white space to make the chart look cleaner and printed material easier to read. Don't begin printing at the edge of the paper and try to keep a margin of three to four inches. If the points consist of a word or two, a seven or eight inch margin may be appropriate. Whatever margin you leave on the left, try to leave a similar margin on the right.

Chalkboards

The same tips for flipcharts can be used for chalkboards and whiteboards. Some extra differences that will enhance your use of these media are:

- Keep it clean.
- Use colored chalk for emphasis.
- Prepare extensive drawings before session starts, or use handouts or other media.
- Use paper to cover the board until ready.
- Maintain eye contact with the audience. Don't talk to the board.
- Print words neatly and large enough for all the audience to see. Remember the back of the room needs to see.

Whiteboards

- Use water soluble felt-tip markers to draw guidelines.

- Use water-soluble hair spray to protect complex drawings.
- Use colored pens for emphasis, remembering the color combinations given for flipcharts.
- Don't stand in front of what you have written.
- Use templates made from cardboard.
- Outline drawings lightly prior to the session.

VIDEO AND AUDIO CLIPS

These media can enhance your presentation when used correctly. Many materials are already on the market that can serve your purpose. A key point is that you should be familiar with the media before you present it to the audience. The following list should be reviewed before using these media:

- Is the film or video self-contained? Is its message what you want? While much of a video/film may be suitable for your purposes, it does require that you preview and mark just how you will use the footage. If you wish to stop at key points and need to fast-forward or rewind to other parts of the media, consider copying only the segments you need that fit with your presentation. It is poor practice to fast forward or rewind to find your segment. Laserdisks and compact disks are easier to use since they can be easily programmed and reset to find the images and information you need. They require more expensive equipment, however.
- Preset the equipment for volume, color, and focus. Set the media where you want to start. It is also good to prepare for likely problems that may occur.
- For TV or monitor viewing, you need to be certain that all members of the audience can easily see the screen. A rule is that there should be one inch of screen per person in the audience (e.g., a 19" TV monitor is good for about 19 people gathered near the monitor).
- Since you should know the program, you can either watch the screen with them or watch the audience to see how they are reacting to the media. If the audience is becoming bored, perhaps too much information is being presented.

MODELS/REAL ITEMS

If you are talking about the details of a complex object, such as a proposed visitor center or a tree branch then a model or the real item can be an incredible benefit to get your audience to understand your ideas. Nothing quite substitutes for "seeing" the real object or a scale/representation of it. If a model or real item can be broken down easily, then it offers opportunities for the audience to learn the intricacies of the topic that words themselves would be hard to convey.

When using real objects or models, be certain that the entire audience can see them. Too often, only the front row of the audience can easily see a displayed object, causing the rest of the audience to lose interest. Even passing the object around the room may not solve this problem, since the speaker may be discussing another point by the time some audience members receive it.

WARM FUZZIES

Actors in Hollywood always warn of the dangers of working with animals. They tend to upstage you and become the center of attention. If you are talking about an animal and using it to indicate certain features and aspects of its behavior, it is wonderful as a graphic. But, as soon as you depart from needing the animal as a focus it should be put "backstage" or out of sight of the audience. Otherwise, the audience may just keep watching the animal, which distracts them from your message.

A real story of such animal antics comes from a raptor center in the Midwest. The speaker had a barn owl that had been blinded in a road accident. While the talk progressed, it was wonderful to observe the various aspects of the owl. But, when the speaker began to talk about other topics such as duties of the raptor center, the owl remained on the speaker's arm. Nearly everyone was focused on the bird, watching how it kept turning is head in response to noises from outside the room. They had all stopped listening. Then when the bird finally ejected feces onto the floor, the room erupted in surprised laughter. Yes, it was entertaining, but nobody heard the message about the raptor center. Whether in an interpretative talk at a park or in a more formal setting, beware of the warm and fuzzy appeal that animals have on an audience.

Another speaker in Montana was talking about wolves. The speaker had both a large dog as well as a captive-reared wolf. The speaker compared and contrasted the two animals in an entertaining and informative way. When the talk moved to more general topics, the speaker had an assistant take both animals out of sight. The audience now had the visual aid framework to understand the speaker and keep listening. As the speaker talked about an experience of meeting wolves in the wild, the audience was not distracted by the unpredictable antics of the animals, but instead remained focused on the story telling of the speaker.

Note that in both cases, the speaker was an expert with excellent speaking skills. In the case of the owl, the speaker was upstaged by the bird. The speaker with the wolf, on the other hand, kept complete control of the situation and hence the attention of the audience.

COMPUTER-GENERATED IMAGES AND PROGRAMS

Electronic presentations follow the same rules for use of color, font sizes, and simplicity of content as any other visual aids. It is easy to go "overboard" when faced with endless options provided by computers. The advent of computerized graphics has made presentations much more impressive. Still, it should be remembered that not all locations will have the facilities to use computerized graphics. Even if you do a computer-based presentation, remember the following admonition: "If it can go wrong, it will."

The following humorous experiences will conspire to disrupt your presentations:

- If you don't take your own computer and projection system (two big heavy trunks) your file will not load onto the host systems.
- Compatibility is a myth. Every piece of electronic equipment acts as though it is a singular prototype—even with the same manufacturer's name on them.

- Telephone line link-ups work perfectly until two minutes into the presentation, when they unexpectedly disconnect.

- Brand-new equipment is always suspect. Consider it guilty until proved useful and reliable.

- The host system has a newer version of the program than you do, but it identifies your file and loads it. However, all program updates are so changed from the previous versions that you need a day just to work out the changes. The crucial icons you use are now buried in another menu.

- The host system has a lower version than you do and thinks your file is just a mass of funny faces, geometric shapes, and odd squiggles.

- You have a Mac and they have a PC that keeps wanting to erase and reformat your disk! Or, vice versa.

- Room service found your disk and pinned it to the room refrigerator with a complimentary hotel magnet.

- Projection equipment light bulbs work for years until you begin your presentation.

- Locking rings are never locked when you come to work with them.

- Your cables are incompatible with the host equipment.

Even if you anticipate everything you can, you can still fall prey to the electronic gremlins. One speaker from the United States went to China to do a presentation. After talking by phone and e-mail to the hosts, who were proud of their new electronic theater set-up, the speaker was assured that all that the speaker had to bring was the file disk with the presentation saved on it. The speaker saved the file in four different formats and versions of the program. There was not just one disk copy, but three, and they were kept in two different pieces of luggage and one on person. The speaker even memorized the layout of the main program icons in case the computer was using Chinese icons. When the day for the talk arrived, one of the versions of the file loaded fine on the computer in the Chinese theater projection booth. There was even a remote "mouse" for controlling the computer from the front stage. The computer was using Chinese, but the speaker was able to "click" the correct icons to control the file during the presentation. Then, when an obvious error message suddenly came up, the speaker had to guess what was written and just click out the message box. The message box disappeared and so did the whole file, and the computer shut down! Fortunately the speaker here had backed up all the presentations onto overhead transparencies and was even ready to talk just using a white board that was in the room.

Whenever you are using electronic equipment, be prepared for all contingencies. One of the first major decisions is whether to take your own equipment, which means carrying quite bulky cases, or to rely on the unknown equipment at the presentation location. Calling ahead to the location of your presentation and getting the exact details of the equipment is the best you can do for yourself to avoid unforseen problems. Always remember to:

- Have spare bulbs
- Have extension cords
- Have 3-prong adapters

- Turn the computer on, turn *down* lights (avoid dark). When finished, turn *on* lights then turn off the computer
- Practice so there is no "dead time"
- Have a contingency plan . . .

CONCLUSION

Visual aids are a valuable method to enhance your presentation. But remember, you and your message are the central components of your presentation. Computer-based technology is likely to become easier to use and more widely available, and will become as commonplace as overhead projectors. Whatever the future trends of graphic enhancements, remember that is all they are—enhancements to your message. Things will go wrong and equipment and power will fail, often at the most inappropriate times. Don't be discouraged. It happens to everyone, and your audiences will be sympathetic. What will set you apart as a professional is how you handle the situation. Be prepared. If everything electronic should fail, then plan how you might use whatever other resources are available. The simpler graphic aids such as flipcharts, chalkboards and whiteboards do not need power and allow you to build your presentation. Consider them all and become competent in their use.

Box 1: Techniques for Successful Use of Transparencies

Place screen to the side or front corner of room if possible and face the audience. This gives you more flexibility to manage the room and help the audience see the overheads. It also helps create a more relaxed atmosphere if the presenter wishes to move away from the projector occasionally. Stand beside the projector facing the audience so that all transparencies can be read by you looking at the projector and the audience looking at the screen.

Use pointer, or use pencil as pointer (not your finger) over the actual overhead while it is on the projector. Be aware if you are blocking the screen to any of the audience.

Use "disclosure technique" to keep audience focused. Don't reveal all the information at once. Place a sheet of paper over the overhead and move it down as you speak revealing just the section you are presenting at that moment. Cover the screen before removing one overhead and replacing it with another to prevent distracting light glares.

Speak with more volume than you normally use. The listener's attention is divided, and more volume is needed to hold their attention.

Use color to add life to the transparencies. When using colored transparency film, be careful to check that the overhead is readable on a projector. Some colors can wash out and do not project well. Use overheads creatively, e.g., multiple overlays, colored highlights, changing images, to maintain audience interest.

Check legibility of overheads from the back of the room. Don't use only capital letters or all italics. Avoid transparency sheets that are dirty. Try to avoid distracting light leaks.

Don't overload your overheads. A reproduction of a typewritten page is one of the worst transparencies there is. Think in terms of bullets and single words or short statements.

The blank space between lines should be 1½ times the letter height.

Transparencies of forms are useful to demonstrate how to complete the form. You can write on the sheet with water-based pens. But, be careful of doing this on inkjet transparency sheet because it will not wash off. Use another blank water washable film over the inkjet film if you need to reuse the inkjet transparency.

When not in use, turn the projector light off. Turn it back on when you need it later in the presentation.

Box 2: How to Use Slides

- Ensure the visual aids support the message. Say dog, see dog.

- Try to arrange it so that all the slides project horizontally. Vertically shot slides may lose the top and bottom of the slide when it is projected onto the screen.

- Arrange the projector so that the image fills the screen. Move the projector back or use a zoom lens if need be.

- Focus before you begin on a slide you know has a sharp image, and the majority of the slides will be in focus. Use plastic mounts if possible. They stay straight and store easier than cardboard mounts.

- Ensure graphics and photographs are consistent and of high quality. Sometimes you may have to use a poor quality picture copy as with historic photos. Still, there should be no need to use poor "modern" slides.

- Rehearse your presentation with the slides before you give it to the audience. You should have an idea of what is coming next without having to see it. Check that they are all in order, projecting the right way round and the right side up. A simple rule when loading slides (especially in Kodak-type carousels) is to look at the slide as you want to view it and then just turn it upside down and load it.

- Is the slide a plus or a detraction to your presentation? If you show a slide that has a "wow" factor, staying silent for a few seconds will help the audience to absorb the image. Once they are relaxed again you can readily continue with the presentation knowing that they are focused. Be sure that you are talking about obvious points in the slide.

- If you need to point at something in a slide, use a laser pointer. An extendable pointer is still useful, but do not touch the screen since this will create a disturbing rippling on the screen.

REFERENCES AND FURTHER READING

Hager, P. J. & Scheiber, H. J. (1997). *Designing & delivering scientific, technical, and managerial presentations.* John Wiley & Sons.

Ham, S. H. (1992). *Environmental interpretation,* North American Press.

Harris, R. L. (1997). *Information graphics: A comprehensive illustrated reference: Visual aid tools for analyzing, managing, and communicating.* Management Graphics.

Hooper, J. K. (1997). *Effective slide presentations: A practical guide to more powerful presentations.* Fulcrum Publishers.

Kearney, L. & Wilder, C. (1996). *Graphics for presenters: Getting your ideas across.* Crisp Publications.

Leech, T. (1992). *How to prepare, stage, & deliver winning presentations* (2nd ed.). AMACOM.

Morrisey, G. L., Sechrest, T. L., & Warman, W. B. (1997). *Loud and clear: How to prepare and deliver effective business and technical presentations* (4th ed.). Perseus Press.

Rozakis, L. E. (1999). *The complete idiot's guide to public speaking.* McMillan Distributers.

16

DEALING WITH THE NEWS MEDIA

This chapter is meant to help the communicator understand how the news media work, and to give advice on interacting with reporters and producers so that fair and accurate information is reported to the public. There are many misconceptions about the news media and the role they serve in society. It is often assumed that the news media give a complete picture of current events or situations. While news reporters do aspire for completeness, the news process has many limitations that restrict how much news can be reported and how detailed such reporting can be.

WHAT IS THE NEWS PROCESS?

The news media are not filled with an altruistic giving of information by groups to just inform the public. News is a business, one where there is much competition for the reader, listener, or viewer to use specific sources of information. Because of this fierce competition, each news source needs to be able to offer information that attracts and holds a core loyal audience. Most news outlets do not make money selling the information and rely on the advertising that accompanies the news. Look at any newspaper or view the TV news and notice how much advertising is present.

The news media can be viewed as a process by which accounts of current events are transmitted to the mass public. News is information, but that information is controlled such that the news media control what, where, and how we know what is happening in the world. News is event-driven so that it must have some novelty to it. Everyday occurrences lack this. News worthiness depends upon the cultural background of both the editor/writer and that of the intended audience.

THE ROLE OF THE MEDIA

While the news media strive to remain fair and accurate, numerous factors interact to influence how people think and react to the information that is presented. Reporting can make a difference by stimulating action by highlighting specific events and problems. The media act as "gatekeepers" of information in that they control and interpret what we read and see, and also make economic, social and political inferences about the information for the receivers. Media news coverage is a net, not a blanket. Only big stories get caught. There is only so much space or time in which to cover information in the news outlets. In the broadcast news, information is limited to news holes where time constraint is a major factor of what news is transmitted. The news media sell information, not supply it free of charge. To get people to "buy" news coverage means using a marketing approach. Each news media outlet has to entice "customers" to buy their package of news over a competitor's. There are complex issues to be explained for an audience not prepared with background knowledge. Consequently, environmental and scientific issues can be: (1) Distorted—the information is altered to make it more understandable or appealing to an audience that would not otherwise be interested in it; (2) Sensationalized—a personal angle may be given to the story that eclipses the science information that prompted the story; (3) Over Simplified—the science in the story is too complex for the audience to understand, and in simplifying it the information becomes erroneous and even misleading; and (4) Inaccurate because of news reporting constraints. Facts presented are incomplete or fail to give adequate context for understanding the reality they purport to describe.

NEWS REPORTING CONSTRAINTS

Reporters aspire to be fair and accurate. Yet consider the constraints of working for a system that needs fresh stories or new perspectives on older stories as soon as possible after they occur. Old news is not "news," it doesn't sell. The list below highlights some of the main constraints for journalists in reporting timely news (Freidman 1983).

- **The "Scoop"**—this is the one story that is hot news and for a short time is of high appeal to the audience. It may also be limited to a small number of news outlets who can control the story for a short time and increase the audience's use of specific news media choices.

- **Short Deadlines**—There is often a lack of time for in-depth research or source checks to confirm information, so confirmations may not be always reliable as desired.

- **Editorial and Advertiser Pressure**—Biases from public agencies, private experts, and reporters may enter into news reports. Since the news media rely on advertisers to make money, the news being reported must meet certain expectations or the patronage may be withdrawn. Likewise, the editor is responsible to superiors to maintain certain expectations of information coverage.

- **Lack of Consistent Sources/Knowledge**—Complex technical information is the bane of the news media. When "experts" are in conflict over findings, it becomes difficult for the reporters to judge just which views are more correct. It should be mentioned that this is at the heart of the scientific process where research information is debated and facts are never really proved. Most reporters are also not trained scientists, so they rely on experts to put the information into non-technical context for them—something the experts are not always very good at doing.

- **Crisis Orientation**—This is instant sensationalism brought about by some current event which can then be used to maintain a high profile story over a longer period. Examples of this might be the Yellowstone forest fires, the explosion of Mt. St. Helens, or the wrecking of the Exxon Valdez in Prince William Sound.

OTHER LIMITATIONS TO SCIENCE AND ENVIRONMENTAL REPORTING

In reporting a regular story, the reporter can ask the following: Who? What? Why? Where? When? and How? In science and environmental reporting, the uncertainty of these six aspects can create many problems with the story.

- The news story is believed accurate by the public, although potentially it may not be.

- Environmental issues are complex; there is no single problem nor is there a single solution; solutions often involve trade-offs, yet this complexity is often disregarded to make the story usable and more understandable.

- Environmental issues often occur because of long-term cumulative effects of the issue in which it is difficult to find a central focus to anchor the story.

- Since environmental issues are so complex, they often get "dove-tailed" with other social concerns. Thus, by linking them with something simpler and more understandable, this dilutes the issue and reduces the need to explain it more.

- Environmental reporting is interpretive since it has many aspects to it that are uncertain. There may be a reliance on anecdotes to help it become more explainable.

- Complaints about what is important in a complex environmental or science story frequently stem from varying perceptions held by varied audience members.

- Scientific language and symbols, which may have specific meanings in an environmental/science context, are often used in general ways in news stories that may cause errors of understanding. For instance, the statement used by the news media "vaporized as in an atom blast" might have been stated by a scientist as "carbonized," "combusted," "or rapidly oxidized."

- Sources who make the journalists work easier are more likely to get coverage. But, this needs to be recognized as a form of subsidy.

NEWS RELEASES

The media often become aware of a potential news item, especially those that occur within or through organizations, by the use of news releases. The production of these is covered in more depth in the news releases (Chapter 9, #1). While news releases are a useful way to let news editors know about your organization and its activities, it must be stressed to send information that is pertinent to a larger audience. Remember that a news release needs to be newsy, have a "grabber" of interest, and be focused on the audience of the specific news outlet you send it to. News media is written in a highly structured format (inverted pyramid format) that allows the editor to made quick cut of the copy to make it fit into available space in the news outlet. This format is described more in Figure 16.

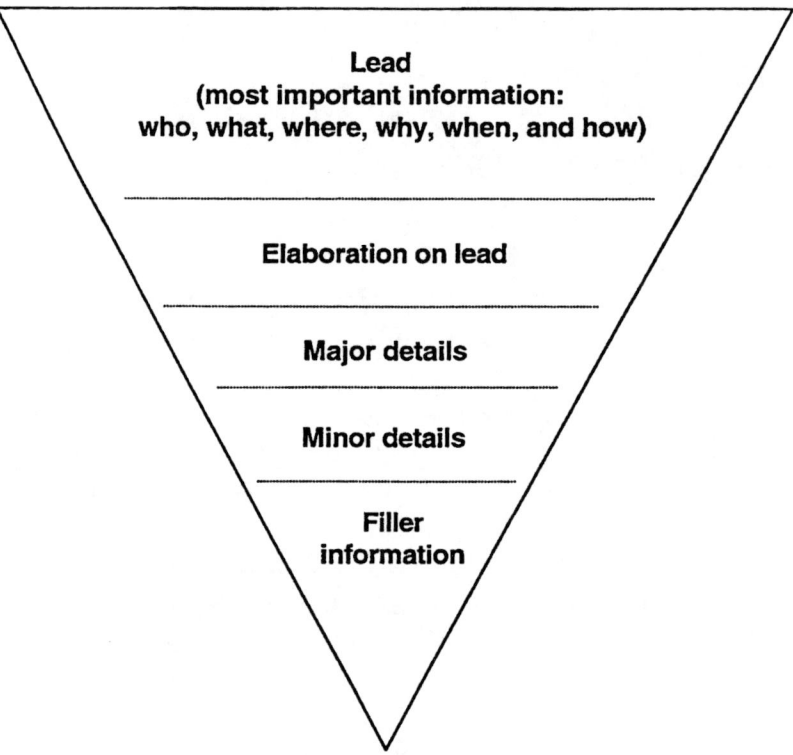

**Lead
(most important information:
who, what, where, why, when, and how)**

Elaboration on lead

Major details

Minor details

**Filler
information**

This is a highly structured format of writing that lets an editor easily chop the length of an article. This allows for easy inclusion into a publication without having to search for the essential information or do a major rewrite of the material. The editor may simply chop a sentence or even whole paragraphs from the bottom upwards to make it fit the available space. Hence, essential information must be at the top of the article. Occasionally an editor may ask for more detail if the story is interesting and space is available, so that the pyramid may be extended with filler information without having to modify the main story already written.

Figure 16   Inverted Pyramid of Writing

NEWS MEDIA OPTIONS

When sending a news release, it is essential to understand the needs of the targeted outlet's audience. The following outlets have different sizes of audience as well as varied special needs.

- Newsletters—this form of news is usually done through organizations to carry information pertinent to specific audiences associated with the organization.

- Community newspapers—usually come out weekly and are provided to each household in a neighborhood. Primary information deals with the neighborhood and surrounding communities.

- Weekly newspapers—many smaller news outlets are similar to community newspapers. They tend to be circulated only once or sometimes twice a week. They also tend to report mainly local information.

- Daily newspapers—published in specific locales and usually circulated in the morning (although some evening papers still exist). Can be delivered, purchased from special curbside dispensers, or bought at newsstands. While some papers are restricted to specific cities in which they are located, some major papers have national distribution from a central city with branch offices in major cities across the nation, to handle local issues. Individual city editors (gatekeepers) often use a wire service for stories which they may use directly as received or edited to fit local needs.

- Wire services—It is important to understand that wire services do not actually print any information, they supply it to the other outlets (newspapers and broadcast news) that do. While some of the larger daily papers will have their own staff reporters, most local newspaper outlets cannot afford to have reporters cover news nationally and internationally and so rely almost exclusively on the wire services to carry news stories from outside their home service area. The services cover what they perceive will be the news for the day and hence that is what is reported by subscribers. News leads in the wire services run in minutes and not hours as may be the case in daily local newspapers.

- Magazines—Some of the news magazines may be a good source for specialty stories involving an organization. Be aware of the needs of the magazine and the audiences they serve.

- Broadcast news media—If using the broadcast outlets for breaking news, the news release is best faxed or even e-mailed to the news director. Phoning them beforehand is also recommended to alert them to the news. If the information is less timely but of interest for a feature, then a news release with information/fact sheets will allow the editor to determine if a reporting team should be sent out to talk with you.

SCIENTISTS/ENGINEERS AND THE NEWS

It is important for scientists and engineers to understand that there is a big difference between what they want to say about their research and what science writers/reporters will actually report! Reporters write to inform, not educate. The gatekeepers do the selecting and the "selling" of the information. Reporters in competition with other reporters have an allegiance to the truth, and are looking for answers that will address the "good" for society.

The Reality of Science Reporting

- Reporters are concerned with the application and social relevance of science, not implications to the science discipline. Hence, scientists and engineers must interact with the news media on these terms. Scientists need the news media in order to promote science outside the science community.

- Scientists who reviewed science media stories generally concluded that one-third of stories had inaccuracies, omissions, or were too vague. Reporters retorted that this one-third represented information of little value to the lay reader. Scientists don't understand audiences and especially the cognitive limitations of the 95% of the public who are not scientists. Journalists understand audiences more than the science.

- Scientists need to relate the research findings with some logical topic of public interest. For example, an article about slime mold flagella might be related to sperm motility.

- The uncertainty of science is not realized. Scientific statements are seen as definitive.

For Scientists to Connect with Reporters

- The information needs to have been published or at least accepted to retain credibility. The classic case of cold fusion in a test tube emphasizes this point. Two scientists contacted the news media and held a press conference about the findings of their research in which they purported to have created cold fusion. The significance was simple, safe and endless cheap energy—a science fiction reality. Unfortunately, the scientists had not published their results in a peer-reviewed journal. When other scientists tried to reproduce the results, they couldn't. The whole experiment was nothing more than a sham (Platt 1998). Yet the public had been alerted about something that was not real and in the public's eyes scientific credibility suffered for all scientists.

- Scientists need to be prepared to address what is important to the reporter:

- What are the ESSENTIAL findings? State facts clearly; limit use of scientific jargon; remember the audience may not know any of the science involved with the issue. What is the societal importance! Present accurate information; do more homework to build a good knowledge base.

- Forget the gory details of methodology and statistical outcomes. If you can't give a definitive answer, think twice about stating it!

- Analogies are really helpful if some methodology must be given.

Understand News Criteria Importance

- Is the information relatively new and is it of reader/listener interest? Address the larger meaning or significance to the story. Environmental issues are complex. Give more than just the basic facts. News must be current or *relate* to something current.

- Is it of potential significance to the reader/listener and unique enough to be reported? Give a local scope to the reader; relate this issue on a local level (the issue or event does not have to have *occurred* locally). Story should have a human interest aspect. Answer the question, "Why do I care?" or "What's in it for me?"

- Will the information interest the writer and the editor?

- Is the information source prominent enough to give it a personal aspect (e.g. a visible scientist)?

- Is the information of intrinsic importance to science?

- Keep it succinct and to the point. If written, provide detail in the correct format (inverted pyramid). If verbal, through radio or TV, have information ready to state clearly and without unnecessary pauses or hesitations.

CONCLUSION

Dealing with the news media is an increasingly important aspect of environmental communication. News coverage can be extremely effective in transferring messages about resource management issues, but it can also be damaging. The environmental communicator must understand that the news process is dynamic, and the quality and quantity of coverage can be influenced. Positive interactions with the news media are most often the result of careful communication planning and a keen understanding of the news process.

REFERENCES AND FURTHER READING

Friedman, S. M. (1983). Environmental reporting: Problem child of the media. *Environment*, 25(10): 24–29.

Gause, V. (1997). *A beginner's guide to media communications*. National Textbook Company.

Jones, C. (1999). *Winning with the news media: A self-defense manual when you're the story*. Video Consultants, Inc.

Mencher, M. (1997). *News reporting and writing*. WCB/McGraw-Hill.

Platt, C. (1998). What if cold fusion is real? *Wired*, 6.11, Nov. 1998. http://www.wired.com/wired/archive/6.11/coldfusion.html.

Rich, C. (2000). *Writing and reporting news: A coaching method* (3rd ed.). Wadsworth Publising Co.

Stephens, M., & Lanson, G. (1997). *Writing and reporting the news* (2nd ed.). HBJ College & School Division.

West, B., Sandman, P. M., & Greenberg, M. (1995). *The reporter's environmental handbook*. Rutgers University Press.

17

MANAGING CONFLICT

The history of natural resources management and environmental issues is filled with notable conflicts. Indeed, in an article about ecosystem management, the Ecological Society of America pointed out that these are not conflicts between humans and nature, but conflicts between competing human needs for natural resources (Christensen et al. 1996). This chapter is not meant to be an in-depth analysis of conflict management, but more a guide to help the communicator understand their role in conflict situations. We have chosen the term "conflict management," rather than "conflict resolution," because many situations that arise in environmental issues are not resolvable. Natural resource managers are best suited to finding ways to make conflicts as resolvable as possible, rather than letting these conflicts debilitate organizations and communities.

Conflicts are usually based in differing values among the disputing parties. Five bases of human values can be attributed to ecosystems: 1) economic output (making money from natural resources), 2) ecological services (human needs such as clean water and climate regulation), 3) aesthetics (beauty) and spirituality, 4) ethics (the moral obligation to protect the natural environment), and 5) education (the scientific and learning aspects of the environment) (Mullins 1996). Nearly all people hold all five values related to these bases, but prioritize them differently. People are reluctant to risk the value(s) that they prioritize highest, thus the rancher may value the educational aspects of a rangeland, but seeks first to protect its economic value for livestock grazing. By contrast, the recreational hiker may be unwilling to compromise on the aesthetic/spiritual value of a wilderness area, even though they support the idea of scientific research in the area. These differences in values' priority are the root of conflict.

In cases where disputants in a conflict can each achieve protection of the values they hold highest, we call the solution "win-win." In contrast, when one side of a dispute achieves their goal, while the other side is unable to obtain a satisfactory solution, we have a "win-lose" outcome. All too often, environmental conflicts are expensive, drawn out situations in which all involved fail to achieve an acceptable outcome. These are called "lose-lose." Skillful conflict management is essential to achieving win-win outcomes to natural resource debates.

REASONS FOR CONFLICT

Conflict is arguably an inherent part of human society, and is not necessarily bad. For example, conflict prevents stagnation, stimulates creativity, allows disputes to be aired, provides a forum for testing ideas, creates cohesion in a group, and can be a source of energy. The negative effects of conflict, however, are more often considered: wasted resources in competition, creation of misperceptions and biases, decreased communication, blurring of issues, magnification of differences, escalation of conflict, and locking into positions (Carpenter & Kennedy 1988). The goal of communicators involved in conflicts is to take advantage of the positive aspects of conflict, while minimizing the dysfunctional effects.

Conflict can occur at any scale. It can be found between individuals, among members of a group, or among different groups. Indeed, in complex environmental issues, all of these types of conflicts may occur simultaneously. Different levels of conflict are observable in most issue/environmental disputes and conflict problems. While a war or fight metaphor is often used in the words describing conflicts, the actual tactics used in a real issue may be more psychological than physical. But, such abuse and disregard are every bit as damaging. Though many issues or problems go further than the contention stage, the communicator should judge if the conflict has already reached a higher level with some or all of the protagonists (emphasized by hurt, disgust, agitated anger, and possibly hatred). If it has escalated, then an expert may be the only way to help bring the issue back to a manageable level.

ANATOMY OF CONFLICT

As stated earlier, conflicts are created over prioritization of environmental values. As disagreements over these values arise, different behaviors may be expressed by disputants. They may choose to avoid the conflict, which generally only delays it until a later time when there are fewer options for a mutually acceptable resolution. Disputants may choose to collaborate to seek a true win-win solution. Or, more often, they may enter into a competition, seeking to win the conflict regardless of the impact on the other party. Sometimes, disputants become aggressive, seeking not only to win the conflict, but to inflict political, psychological, or even physical damage on the other party.

Unmanaged conflict develops in fairly predictable ways. Figure 17 illustrates a typical social conflict (Lewicki et al. 1994). The spiral depicts the many aspects of an escalating conflict. Notice how the two separate groups in the left columns treat the conflict situation as it develops. The right two columns emphasize how the parties begin to take on aspects that were not part of the original problem. Note also in the last column how the mind-set changes quite drastically from one of concern to one of a power struggle. Eventually aggression against opponents, and the elevated conflict level themselves become the issue that seems to consume the participants' energies. At the top of the spiral, it seems winning is more important than resolving the underlying issue. It should not be construed that the spiral indicates a crescendo situation with no resolution. Rather, it empha-

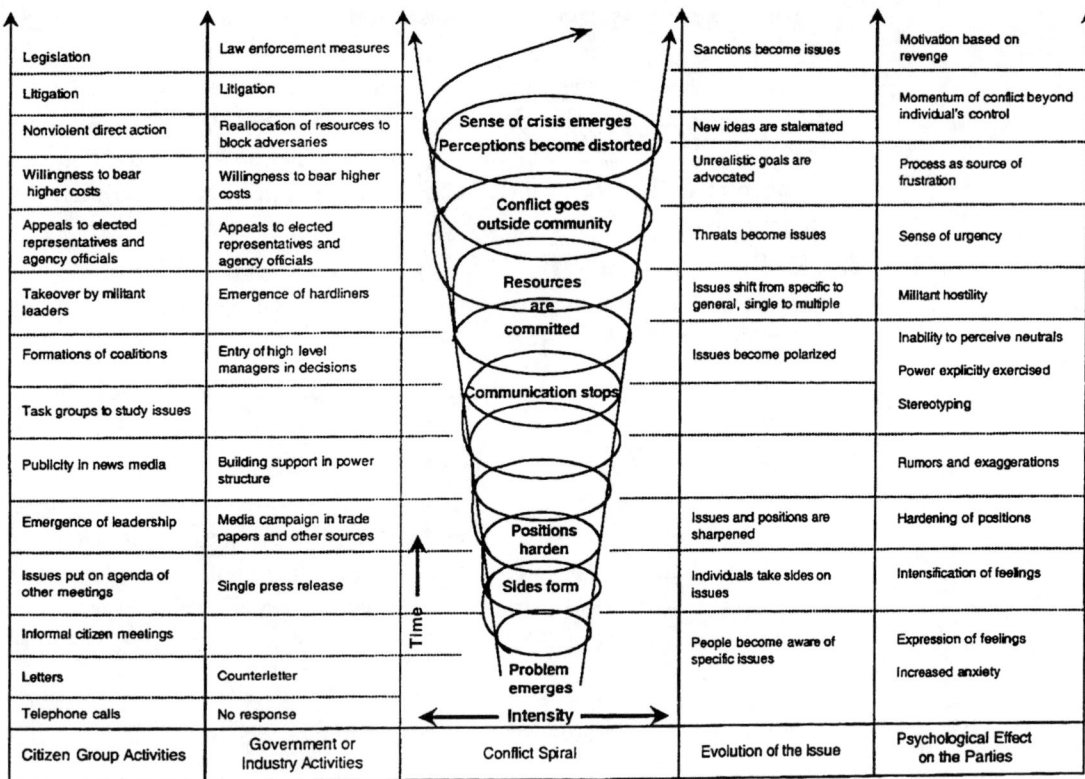

The spiral (Conflict Spiral column, bottom to top) reads: Problem emerges — Sides form — Positions harden — Communication stops — Resources are committed — Conflict goes outside community — Sense of crisis emerges / Perceptions become distorted. Axes labeled Time (vertical) and Intensity (horizontal).

Citizen Group Activities	Government or Industry Activities	Evolution of the Issue	Psychological Effect on the Parties
Legislation	Law enforcement measures	Sanctions become issues	Motivation based on revenge
Litigation	Litigation		Momentum of conflict beyond individual's control
Nonviolent direct action	Reallocation of resources to block adversaries	New ideas are stalemated	
Willingness to bear higher costs	Willingness to bear higher costs	Unrealistic goals are advocated	Process as source of frustration
Appeals to elected representatives and agency officials	Appeals to elected representatives and agency officials	Threats become issues	Sense of urgency
Takeover by militant leaders	Emergence of hardliners	Issues shift from specific to general, single to multiple	Militant hostility
Formations of coalitions	Entry of high level managers in decisions	Issues become polarized	Inability to perceive neutrals / Power explicitly exercised
Task groups to study issues			Stereotyping
Publicity in news media	Building support in power structure		Rumors and exaggerations
Emergence of leadership	Media campaign in trade papers and other sources	Issues and positions are sharpened	Hardening of positions
Issues put on agenda of other meetings	Single press release	Individuals take sides on issues	Intensification of feelings
Informal citizen meetings		People become aware of specific issues	Expression of feelings / Increased anxiety
Letters	Counterletter		
Telephone calls	No response		

Figure 17   Spiral of unmanaged conflict

sizes that unless expedient action is taken to understand and manage the conflict it can become uncontrollable.

The spiral suggests one way to constructively manage conflict. Disputants who emphasize interests, as opposed to positions, are more likely to find a mutually acceptable solution to their conflict. This is best achieved through principled negotiation. If a dispute can be structured as win-win, it uses a constructive problem-solving approach. This approach of successful resolution is to:

- Understand and act on the interests and not the positions of the parties involved.
- Separate people and their feelings from the problem.
- Develop options for mutual gain of all the parties involved.
- Insist on fair and accurate criteria (not emotional-subjective ideas).

Once positions of where people stand on the situation are defined, it is much more likely that either one or both parties will have to accept a less-than-desirable outcome. The following section describes the alternatives that disputants have for resolving conflicting positions.

RESOLVING DISPUTES

Toward the top of the spiral, disputants will be unable to find a mutually acceptable solution to their conflict. When this occurs, they have limited options for ending the conflict. The major options are described below.

1. **Negotiation**—Parties to the conflict work out a resolution between them. Find common ground and collaborate on a consensual resolution. This

method negotiates something reasonable outside the formal mechanism and can result in a win-win situation. It may require a peer mediator or process manager to keep the negotiation on track, because it requires process skills and an understanding that the other party's values are legitimate. Accepting the values of both sides (or more) in negotiation is possible if the two value frames overlap (Lewicki et al. 1994). More often, however, compromise solutions are developed that require both parties to give up something.

2. **Third Parties**—Ideally, get an **uninvolved third party**. All people with a stake in the issue must agree beforehand on the type and choice of who will be the mediator or arbitrator).

   (i) **Mediation**—Sometimes the mediator facilitates a face-to-face dialogue between the parties. Other times, the mediator is an intermediary, travelling (shuttling) between the parties.

   (ii) **Arbitration**—A third party reviews all the evidence and then makes a decision which may be binding (legal) or non-binding (voluntary acceptance). This must be agreed upon before arbitration begins. Legal contracts may need to be drafted beforehand to ensure that compliance is followed in event of a decision that is not fully acceptable to one of the parties.

3. **Legal remedies**—Resolutions are determined by officials in the legislative or executive public sector.

   (i) **Litigation**—The issue enters the legal system. Court systems are monetarily expensive and prone to extensive delays, leaving many options for countersuits. Having "standing to sue" equals being "qualified to sue" by virtue of being affected by the issue. There is also a lack of expertise by the judges who rule on procedural issues and not the content of the issue. It is nearly always an adversarial situation resulting in a win/lose outcome.

   (ii) **Legislation**—This is lobbying for political action in the form of laws. Legislative involvement by its nature is expensive and requires lobbying skills. Access to legislators is limited. Like the court systems, it can be expensive and prone to delays. Assuming proposed bills make it to a legislative floor, there can still be deadlock on controversial issues. Legislative action can be pursued at the local, state or federal level. However, at the federal level, site-specific local issues may not be understood and the voices of minority groups may not be heard.

What should be obvious is the increasing amount of time and resources that are invested in the management process to resolve the issue that has circled up the spiral of conflict. The sooner the conflict is managed and resolved, the less money will be spent and resources used, and the more likely a suitable win/win result will occur.

COMMUNICATING ABOUT CONFLICT

Carpenter and Kennedy (1988) offered the following guidelines for resolving conflicts. These are also appropriate for communicators to consider in planning messages for disputants. Messages are powerful in the charged atmosphere of a conflict. The communicator has an obligation to not further enflame the dispute. All communications can be crafted around these guidelines.

- Create a supportive climate in which all parties feel as though their input is acceptable.

- Use descriptive speech and thinking. Do not evaluate. Language is the tool we use to resolve conflict although its misuse can create or elevate conflict.

- Be flexible and spontaneous. Have a problem-solving attitude, not a combative one.

- Have empathy, not sympathy. Truly try to understand what are the other points of view and positions. Understand the problem, in all of its dimensions.

- Demonstrate fairness. Do not be hegemonic.

- Realize that conflicts are not just about substance, but also about procedure and relationships.

- Plan a management strategy, then follow it.

- Build positive working relationships between the disputants (and facilitators, if applicable).

- Begin with a constructive definition of the problem.

- All parties should help design the management process and solution.

- Good solutions are based on interests, not positions.

- Be flexible and have a problem-solving attitude.

- Think through what may go wrong. Be prepared.

- Above all, do no harm. You may fail the first time, or have to work with these parties again. You want to garner their respect even if you disagree on principles.

IDEAS TO PONDER

The best way to deal with conflict is to recognize that it is inevitable and to manage it at the outset before it can escalate into something beyond control. Consider the following:

- What are the reasons for governments to exist? They are institutional arrangements related to legal remedies. The government provides services and manages conflict. It is also for the authoritative allocation of values (e.g., Congress passes acts such as the Endangered Species Act).

- Disputes occur over how landowners can use their land. We have social institutions to resolve land use conflicts. Typical conflicts are between and among private users, and between and among public and private users. What the public often wants and what the private owners want will be different. Finite land resources and a growing population will inevitably lead to differing agendas on the use of land.

- There are conflicts between the haves and the have-nots. Those that "have" the resources tend to not share equitably with those lacking the resources. Examples are water rights, public land grazing rights, and public land mining rights.

- Some people want to secure resources for the future while others see technological solutions as the answer. There are differing values on how a solution can be achieved. Many people see the need to conserve resources and reduce the impact of existing problems now, while others see new technologies as panaceas for all present and future problems.

CONCLUSION

Conflict is a given in most of today's society and by its very nature, change in society is created by conflict. Understanding just how conflict arises can help the communicator work to minimize or resolve a conflict situation before it escalates out of control. The aim is to foster beneficial changes to society. By working to produce win-win situations in which all parties can gain something of value, conflict in itself is a valuable tool. It becomes destructive when it is ignored or when hegemonic ideas are imposed on others. By involving all parties in the management process and understanding the values that are inherent in the process, conflict management can be a positive process.

REFERENCES AND FURTHER READING

Borisoff, D., & Victor, D. A. (1997). *Conflict management: A communication skills approach* (2nd ed.). Allyn & Bacon.

Carpenter, S. L., & Kennedy, W. J. D. (1988). *Managing public disputes.* Jossy-Bass.

Christensen, N. L., Bartuska, A. M., Brown, J. H., Carpenter, S. D., Antonio, C., Francis R., Franklin, J. F., Machahon, J. A., Noss, R. F., Parson, D. J., Peterson, C. H., Turner, M. G., & Woodmansee, R. G. (1996). The report of the Ecological Society of America on the scientific basis for ecosystem management. *Ecological Applications,* 6(3), 665–691.

Costantino, C. A., & Merchant, C. S. (1995). *Designing conflict management systems: A guide to creating productive and healthy organizations.* Jossey-Bass Publishers.

Crowfoot, J. E., & Wondolleck, J. M. (1990). *Environmental disputes: Community involvement in conflict resolution.* Island Press.

Fisher, R., Ury, W., & Patton, B. (Eds.). (1991). *Getting to yes: Negotiating agreement without giving in.* Penguin.

Kriesberg, L. (1998). *Constructive conflicts: From escalation to eesolution.* Rowman & Littlefield.

Lewicki, R. J., Litterer, J. A., Minton, J. W., & Saunders, D. M. (1994). *Negotiation* (2nd ed.). IRWIN: Boston, MA.

Mullins, G. W., & Watson, M. D. (1996). Developing public education packages: A U.S. National Park Service perspective. In Szaro, R. & Johnson, D. W. (Eds.), *Managed landscapes: Theory and practice* (pp. 593–604). Oxford Press.

Weeks, D. (1994). *The eight essential steps to conflict resolution: Preserving relationships at work, at home, and in the community.* Putnam Publishing Group.

18

COMMUNICATING ABOUT RISK

Ask a scientist what "risk" means and, more likely than not, they will emphasize the likelihood of something nasty happening. But, as communicators and
social scientists have examined the interface of technology and society more closely
during the last 25 years, they have reached a general consensus that non-scientists view risk in an almost antithetically different way from those in the scientific community. Social scientist Peter Sandman (1991) has said, "The things that
kill people and the things that scare people are diametrically opposed."

For scientists and engineers, and those who communicate about their work,
the take-home lesson is this: citing the probability of something negative happening is never enough. If you want people who do not share your scientific viewpoint to accept your information, you'll have to do more than cite probability.

Scientists and engineers who deal in risk are usually driven by numbers and
hard data obtained from tests and experiments. While these may be quantita-
tively valid for setting exposure levels and probabilities of accidents, such scientific observations do not deal with the broader public's hard-to-measure
psychological reactions to risks. The non-probabilistic aspects of risk are the focus
of this chapter.

When communicating about risk, public trust and the communicator's credibility are the most important factors. Content has been shown to be secondary
to these intangibles.

A convenient way of looking at the scientific and non-scientific parts of risk
are offered by the following equation (Hance et al. 1990; Sandman 1991):

Risk = Hazard + Outrage

> where **Hazard = Probability x Consequence**
> and
> **Outrage = the cultural reaction to risk**

Hazard is the specific scientific determination of how harmful a particular risk has been measured to be. It is the likelihood of a problem arising from a specific situation, and how problems will manifest themselves should a problem occur. **Probability** is the statistical likelihood that a problem may arise. **Consequence** is the predicted outcome should the problem become real. And **Outrage** is the perception (real or imagined) of a problem.

WHAT IS HAZARD?

When experts determine risk, the process takes place in the scientific disciplines of epidemiology and risk analysis. Epidemiology is a medical science examining the incidence and distribution of human diseases. Risk analysis deals more broadly with risk from a variety of methodological perspectives. The Society for Risk Analysis, a worldwide organization devoted to the study of risk, defines "risk analysis" as (REF):

> a detailed examination including risk assessment, risk evaluation, and risk management alternatives, performed to understand the nature of unwanted, negative consequences to human life, health, property, or the environment; an analytical process to provide information regarding undesirable events; the process of quantification of the probabilities and expected consequences for identified risks.

Risk is measured through analogy and mathematical extrapolation from research studies often using models (based on animals, plants, and computers), and then by indirect comparison to the human population. One of the biggest problems with risk communication is not the quantification of hazards, but explaining the judgments made by risk analyzers as to what is hazardous and unhealthy and what is not.

Hazard is determined by risk assessment which is the actual quantification of the hazard. How toxic and dangerous something might be. Since most of the testing does not involve humans, some essential information is nearly always missing. With somewhere between 500–1000 new chemicals being introduced to the marketplace per year, this situation is not going to change. There is not enough time, money or personnel to ascertain the toxic potential of even a fraction of these chemicals. You may be surprised to find out that U.S. federal law does not require that chemicals have any risk assessment before they are used.

Risk management is a firm attempt to control hazards in a positive manner. It uses risk communication and quality management procedures to actively control potential exposure to hazards. But when communicators deal with risk, they are usually dealing with audiences who have perceptions of risk far different than those of the experts who quantify the hazards. This component is outrage.

OUTRAGE

Citizens and activist groups do not like hazard numbers any more than industry and government officials like outrage statements not based on data produced via the scientific method. There is built-in tension in any discourse about risky things. What a risk communicator has to learn is that perceptions are tantamount to the truth to the people holding the perceptions. In one's mind, perceptions are reality. We all believe what we want to believe, but base our views of the truth on different sources of facts. Not everyone finds science to be the best or only arbiter of truth. Regardless of the scientific information involved, much of the time an audience does not understand this hard information. They may not wish to understand it, either. They may even feel alienated by it. But, communicating about risk may not be as hopeless as those assertions make it seem. There are factors that play significant roles in the perception of risk.

When experts examine a risk, they focus almost exclusively on mortality and morbidity: how many people die and how many people get sick. But, citizens define risk much more broadly, considering a wide range of "outrage factors" in determining how risky they consider a situation to be. Most always, outrage is not a misperception of the technical data. Rather, it is how people "feel" about a potential or actual problem in their minds. There might not even be a demonstrable problem—no one has died and no one has gotten sick. But, the people are still not happy about the situation. That some group perceives a problem is reason enough to deal with it as a real problem. Perceptions are real. Scientists and engineers need to understand this and deal with outraged citizens as not delusional. This promotes a better society that deals with the risks of modern living.

RISK ACCEPTANCE

So, working to keep outrage low is as much a part of risk management as working to prevent mortality and morbidity. Here are some factors involved in risk acceptance:

- **Choice**—Do the people have a choice on whether they get exposed to potential risks? Sandman notes (Sandman video), "The right to say 'No' makes saying 'Maybe' a lot easier." When people are part of the decision-making process they tend to accept risks more readily. This gives them some control over the risk in their lives. If any risk is imposed without consent, expect negative reactions. Similarly, if people are comfortable with an existing risk, it is more acceptable even if the hazard gradually rises for some reason. But, there may be times when you actually want to heighten an audience's concern about a problem to help safeguard them. Many of the federal workplace rules and regulations are mandated specifically for this purpose.

- **Financial burden**—Risks and benefits should be shared equitably. If a chemical company builds a plant and the neighbors bear the entire risk burden, and yet have no chance of seeing any of the profits, they will be outraged. Other examples might be a property-owner whose real estate value is greatly diminished because of being next to a stream that receives effluent from an upstream feedlot, or homeowners who cannot sell their property because they are downwind from the emission stacks of an incinerator.

- **Ecological justice**—This emphasizes the problem of poor or minority people bearing the burden of most of the industrial environmental problems. Such lack of fairness based on racial, ethnic or socio-economic factors leads directly to problems with outrage. Audiences may not feel it is morally fair to be exposed to more potential problems than other more affluent neighborhoods.

- **Acts of God**—People are forgiving of "acts of God," yet extremely unforgiving of human-produced problems. Do not try to minimize problems you have by likening them to acts of God!

- **News media**—Many risks have a heightened outrage because of media images. News media often report and amplify a problem that is rife with conflict and sensational anecdotes, while ignoring more deadly problems, which happen slowly or lead to more chronic consequences. Popular culture through films and television programs uses industrial catastrophes for dramatic fictional portrayals. These entertainment products get linked in the public mind to real possibilities.

- **Trust**—When facing a risk, people are more likely to stay calm if they trust the organization responsible for placing the risk on them. But, trust cannot be built in the face of an impending disaster. Organizational communications have to be proactive to be trustworthy. Better to always be sincere, respectful, and courteous, than to try to build credibility in a crisis. Honesty is always the best policy.

A communicator's job may not always be to reduce outrage. Occasionally, the communicator's role may be to increase outrage. This may be a selected course of action when a risk exists but no one seems bothered about it. For instance, many times workers who work in high-risk occupations have regulations that mandate they wear specific equipment to protect them. If the equipment is bulky or the protocol required for full compliance is time-consuming, these workers may place themselves in harm's way by skirting safety precautions. Such a situation could be exposed by a cognizant communicator, and outrage encouraged. It is essential for the communicator to assess each situation on its own merit.

If outrage exists where there is no real danger, only a false perception of it, then the communicator needs to address the outrage factors. Then education intervention will allow the audience to understand the true nature of the perceived risks. However, if the hazard is real and outrage is low, then the communicator needs to educate for the true risk to help people inadvertently harming themselves. One of the main sources of information used by the public to understand risk is the mass media.

MASS MEDIA PERCEPTIONS

Most people gain their information about what is dangerous from the news. Science is rarely definitive, yet news media often express concern over the uncertainties in scientific data interpretation. "Experts" that disagree on data interpretations are more newsworthy than the data, and fuel doubt and confusion in the non-scientific public. News accounts tend to give a simplified and erroneous view of the complexities of any situation. There are always uncertainties, but the news media have not figured out a way to accurately report them.

HOW DO YOU DEAL WITH UNCERTAINTY AND COMMUNICATE RISK EFFECTIVELY?

Communicating about risk requires planning. The communicator needs to understand the science behind the hazard as well as the level of outrage exhibited. A first step is to first get the facts.

The following list will help the communicator to gather information when trying to understand the hazard aspect of risk.

- Is there enough information to make a valid decision?
- What data are missing?
- What additional data are needed before a sound decision can be made?
- Are assumptions made explicit in the data used for decisions?
- Are the scientific methods and statistics appropriate for the data used?
- Is all the data open for full scrutiny?
- Have all the alternatives been considered?
- Have the criteria for selection or rejection been clearly outlined and explained?

Having garnered the information, it is probable that a lot of uncertainty still exists. Definitive answers are not always available. Yet the risk situation, real or perceived, must be dealt with and resolved if conflict is to be averted. Once an assessment is made of the situation, a plan should be drafted and used to manage the outrage. The goal is to help outraged people feel that their concerns are being truly considered while educating them about the real nature of the hazard. Of course, if the outrage is justifiably high, then the communication should be about how the hazard is being managed.

Risk is best discussed directly. The air of mistrust still lingers thickly around information about risk. Audiences are more skeptical of messages about risk than any other subject.

Use pro-active public relations. Get interested citizens involved early. Deal with the uncertainty and outrage head-on. Never ignore or dismiss outrage as unimportant. Release news regularly. Do not suppress information, even if the final data is not available yet. Use community meetings. Develop an outreach program. Ask your audience about their concerns. Speak plainly using minimal "techno-babble." Explain technical concepts. Be careful not to talk around the risk or to oversimplify it. Be cautious about the comparisons you use.

Develop a positive and orderly atmosphere where everyone can talk openly about risks, both hazards and outrage. Involve the public in decision-making. Listen to the concerns of all parties. Ensure all the potentially effected parties are been identified and invited to take part. Even small groups can become a source of tremendous outrage if overlooked. Strive to create a dialogue. Do not take an authoritarian stance. Give all parties equitable standing in the decision-making process. If the corporation or government entity has veto power, so should the local citizenry.

State the limits of the data's confidence and admit uncertainty. Discuss data quality. Emphasize what has been done about a problem already and what is currently being done. Never cover up information. Never lie. Be honest. Trust and credibility are absolutely essential.

Risks that frighten are not the always the ones that do the damage. While news media reporting in the past may have heightened concern for certain problems and issues, there remain many hazards that the public is mostly ignorant about. The risk communicator's job can be seen as being smart enough to find situations when either hazard or outrage is predominating the risk equation and informing those affected by the risk of the other side.

FINAL THOUGHTS

It has been said that we are safer and healthier than any other period in history. If that is so, why are we more concerned about risk?

The last two centuries have seen the spread of industry and high technology around the planet. More technology, in general, engenders more risk. That is not to say that magnificent benefits—better health, longer life spans, high quality of life, more entertainment opportunities, faster communication and transportation—have not come to modern humanity. Our greater reliance on new technologies has two important effects when considering the overall societal perception of risk. First, older technologies tend to become taken for granted. Machinery on family farms is some of the most hazardous technology around, but you hear few cries to save farmers from being mauled by their own plows and combines. Newer, unfamiliar technology breeds outrage, as people try to come to grips with its meaning in their lives. Think back over the last decade and the rapid development of the Internet. How many urban legends have you heard from cyberspace?

The second factor our broadening technology base has had on our collective perception of risk is a statistical one. As populations have increased and the amount of technology available has risen as well, more accidents—in absolute terms—have been occurring. Even if each type of technology is dramatically safer, having hundreds or thousands more machines around still increases the number of problems that will occur. A one in a million eventuality can seem almost commonplace in a world of six billion. Couple this numerical twist with a widespread belief that technology can be totally under our control and outrage germinates and spreads.

Experience with some sensationalized catastrophic mishaps of technology such as Three Mile Island, Chernobyl, and Bhopal have heightened our collective social concern to the point that anytime a bad thing happens we are ready to believe the worst possible report. We have become over-sensitized to industrial accidents.

A related social phenomenon is the confusion promoted by dueling experts in litigation. An incredible increase in litigation catches more and more of the public's attention. They see hired guns shooting their mouths off, contradicting each other, all the while using the same data. Credibility of science is eroded.

As we have become wealthier and healthier, it is seductive to feel we have more to lose. This fear can transform into paranoia, an obsessive and acute awareness of being surrounded by risks beyond one's control. To expect zero risk is to guarantee burning outrage. Pushing an image of a risk-free life is an open invitation to failure.

Mass mediated information from well-funded special interest groups plays a significant role in the perception of ever-increasing danger. The message creation and delivery systems of interest groups have become more sophisticated. They can continually bring issues, both new and old, to our notice. And, the number of available channels available to spread information has increased exponentially. In short, we increasingly rely on others to tell us the truth. Indeed, it can be argued that we are so well-informed now from this broadening information base, that we become psychically paralyzed by the knowledge we possess. Such well-informed futility is concomitant with a perception of increased risk (Wiebe 1973).

REFERENCES AND FURTHER READING

National Research Council. (1989). *Improving risk communication.* National Academy Press.

Gutteling, J. M., & Wiegman, O. (1996). *Exploring risk communication.* Kluwer Academic.

Hance, B. J., Chess, C., & Sandman, P., M. (1990). *Industry risk communication manual: Improving dialogue with communities.* Lewis Publishers/CRC Press.

Kamrin, M. A., Katz, D. J., & Walter, M. L. (1995). *Reporting on risk: A journalists handbook on environmental risk assessment.* Michigan Sea Grant.

Leviton , L. C., Needleman, C. E., & Shapiro, M. A. (1997). *Confronting public health risks: A decision maker's guide.* Sage Publications.

Leviton , L. C., Needleman, C. E., & Shapiro, M. A. (1998). *Environmental health for all risk assessment and risk communication for national environmental health action plans: Risk assessment and risk communications.* Kluwer Academic Publishers.

Lundgren, R. E., & Mcmakin, A. H. (1998). *Risk communication: A handbook for communicating environmental, safety, and health risks.* Battelle Press.

Powell, D. A., & Leiss, W. (1997). *Mad cows and mother's milk: The perils of poor risk communication.* McGill-Queen's University Press.

Sandman, P. M. (1991). *Risk = hazard + outrage: A formula for effective risk communciation.* American Industrial Hygiene Association Distance Learning: Professional Development Courses and Products, Fairfax, VA.

Slovic, P. (1987). Perception of risk. *Science, 236,* 280–285.

Sadar, A. J., & Shull, M. D. (Eds.). (1999). *Environmental risk communication: Principles and practices for industry.* Lewis Publishers, Inc.

West, B., Sandman, P. M., & Greenberg, M. (1995). *The reporter's environmental handbook.* Rutgers University Press.

Wiebe, G. D. (1973). Mass media and man's relationship to his environment. *Journalism Quarterly, 50,* 426–432, 446.

Wogalter, M., Dejoy, D., & Laughery, K. R. (Eds). (1999). *Warnings and risk communication.* Taylor & Francis.

19

LEARNING FROM MARKETING AND PUBLIC RELATIONS

Marketing and public relations are inventions of the business world for communicating messages to influence opinions of specific audiences. Popular opinions about these fields are often less than flattering, especially by those with environmental leanings. Indeed, a natural resources management student responded to a lecture on marketing communication with the comment that conservation was "much too important to use marketing" in its pursuit. Furthermore, the concept of public relations is often thought of as putting "spin" on bad situations to make them acceptable to the public.

In contrast, we believe that both marketing and public relations represent powerful communication planning processes that can be, and indeed are used to deliver messages to audiences with respect to environmental issues. In this chapter, we will give a brief overview of marketing and public relations and highlight how they can be used in environmental communication. Note that they are both planned communication processes, and are quite similar in process to the communication planning model presented in Chapter 5. We contend public relations and social marketing offer important and useful tools to environmental communication.

MARKETING AND SOCIAL MARKETING

The field of marketing is based on the concept of the market—a place where goods and services are exchanged. In most markets, participants exchange goods, services, money, information, and even time. The concept of exchange, then, is critical to marketing. One can think of marketing as the managerial process by which people, or organizations, share information about what they would like to exchange. Most marketing texts distinguish marketing as a managerial process and a planned activity to understand an audience's needs and wants, and to find a means to fill these desires. A critical aspect of marketing is the concept of discovering unmet needs; marketing is not creating wants for products already in hand. Creating demand for items in hand is selling, not marketing.

A common example of marketing applied to a natural resources communication problem is the concept of membership in a conservation organization. In order to raise money for the organization, they may offer enticements in exchange for your donation. For example, a conservation organization may offer a tote bag, a magazine, or a coffee mug for a certain size donation. To encourage larger contributions, the enticements may increase in value for larger donation sizes. Marketing is more than just having something to exchange. It is a planning approach based around four concepts: product, price, placement, and promotion.

Product is the item to exchange. In the marketing approach, the communicator tries to define for the audience a product that can be consumed. In the example above, organizational membership was defined by small gifts that were exchanged for membership fees. This will rarely generate enough funds for a conservation organization to operate. If the organization can infuse positive feelings of helping the environment into the product, then membership is far more powerful. Then, however, the organization must deliver: the member must be made to feel like they truly have helped.

Other products can be used in marketing approaches. In natural resource management, the resources themselves can be products. So can privileges, such as the ability to hunt on a property, or use of a campsite. The marketer always tries to think of value exchanges by defining consumable products.

Price is the amount the audience is willing to exchange for the product. Price is usually thought of as the amount of money that an audience exchanges for the product. But, price can also be non-monetary. For example, price can be time, bartered goods, opportunity costs, and peripheral costs used in the exchange. When using a marketing approach, the communicator evaluates the entire cost to the audience, not just the amount the "seller" will receive. For example, a community comparing the options of curb side recycling and a centralized recycling center would include the costs of participant travel to the recycling center in the comparison, including gasoline and travel time.

Placement refers to the distribution of the product. The marketer must answer the question "how can the audience obtain the product?" Many environmental groups solicit memberships by mailing promotional materials to perspective members' homes, thus making for convenient placement. The new member need only write a check, or give a credit card number, and mail it back to the organization. As another example, hunting and fishing licenses are often sold by state departments of natural resources through sporting goods stores, which are places that hunters and anglers are likely to visit anyway.

Promotion refers to the manner in which the audience is made aware of the product, as well as its price and placement. Advertising and personal selling are two ways, among others, that this is achieved. Promotion also refers to the message sent about the product: is it healthy, environmentally-friendly, or money-saving? The audience is provided information to make the decision to obtain the product. As an example, some states have tax check-off programs to fund natural areas protection; these are often promoted through posters and special mailings.

Marketing, then, is the application of a planned process to meet the wants and needs of an audience. Social marketing is the application of these principles to the goal of achieving social change (Zaltman et al. 1972). The product becomes the social changes itself, while the price is what individuals give up to achieve the change. Place, as defined by Zaltman et al. (1972, p. 8), is "outlets which permit translation of motivations into actions." This gets at the idea that while people consider themselves environmentalists, they do not have the opportunity to act on their expressed ideals. A marketing approach to environmental behavior must address this need for actionable options. Promotion, in a social marketing perspective, is the raising of awareness of the desired change.

Marketing and social marketing are powerful tools for communicators. When resource managers adopt a product orientation, they may find new, and more effective means of achieving objectives. Certainly traditional natural resources, such as forest products, have been approached through marketing. Many other resource problems can be addressed through product orientation: interpretive services, recreation, fund-raising, and volunteer recruitment. Objectives specifying behavior change, such as recycling, green consumerism, saving energy, eliminating poaching, and so forth can also be addressed using a marketing approach. This is an important component of the environmental communicator's toolbox. At first, using social marketing techniques will feel highly creative. With time and success, such skills become routine as well as valued.

PUBLIC RELATIONS

The field of public relations is often maligned. The phrase can connote communications designed to "spin" unacceptable events to make them palatable to an outraged public. This is probably caused by the use of public relations strategies after some event has occurred to try to mitigate its negative impacts. Many managers only think of public relations reactively, after a public is outraged by some event or action. However, public relations, when used proactively, can be a powerful, two-way communications tool for an organization to establish trust and mutual benefits with its publics.

The Public Relations Society of America uses the following definition: "Public relations helps an organization and its publics adapt mutually to each other" (PRSA 1999). This definition indicates how public relations creates a dialogue between the organization and the community. Indeed, many "public relations disasters" would not occur if sufficient public relations work was done as an ongoing part of business.

Several specific types of public relations are practiced by organizations. Each of these use the same principles and are practiced by all types of organizations, but have different goals and specific activities. All of the types of public relations described below are common in environmental fields.

Media relations consists of establishing and maintaining working contacts with the press, including local and national newspaper, television, radio, and other mass media outlets. In media relations, it is important to understand that credibility and trust must be established at two levels: first with members of the press, and second with the final audience. Relationships can be fostered with reporters and editors through careful preparation of news releases and offering frank, honest interviews, demonstrations, and tours. Exclusive stories are valuable, so offering an interview to a single newspaper or TV station can increase its probability of being used, and can build goodwill with the journalists. It is important when working with the media to understand the needs of reporting, as discussed in Chapter 16.

Government relations are those activities intended to influence policy making (law making, and rule enforcement) by government organizations. Personal actions such as lobbying are often used, but use of media and mass mail campaigns are common supporting or primary activities of government relations. This area requires not only an understanding of the principles of public relations, but also knowledge of how government works and regulations concerning lobbying.

Community relations include activities to include local stakeholders in planning and decision-making. Even when companies, governments, or nonprofit groups have no legal obligation to work with local residents, good community relations can lead to creative problem solving and reduce the likelihood of resistance to actions. Besides being a good neighbor, organizations can practice open, honest dialogue within the community, both listening to concerns and providing meaningful, appropriate information. The principles of risk communication (Chapter 18) and conflict management (Chapter 17) are important in this area of public relations.

Crisis communications are those public relations activities used when disasters occur. Crises can involve corporations, such as the *Exxon Valdez* oil spill, governments, such as the space shuttle Challenger explosion, and even nonprofit organizations, such as the recent United Way scandal. Organizations should prepare crisis communication plans in advance, by determining which kinds of crises might occur, and preparing communication plans in advance, including designating who should be on the communication team. Even the unanticipated crisis should be considered with a generic communication plan. As crises are inevitable, organizational response can save lives, ecosystems, reputations, and money.

Internal communications are those for which the target audiences are employees, members, or volunteers of the organization. These messages are designed to keep internal audiences informed and motivated. Media include printed media such as newsletters, electronic media such as videotaped addresses, personal presentations, "town hall meetings" and one-on-one discussions. Like all public relations, open, honest, two-way communication is critical. It is important to remember that the boundary between the organization and external audiences is permeable: employees hear external messages, and external audiences eventually hear internal messages.

Public relations, according to the PRSA (1999), benefits both the organization and society. It allows the organization to be responsive to its publics, thus making it more effective. At the same time, public relations informs the organization about its social responsibility by giving voice to community constituencies. The key to effective, meaningful public relations is the same as other communication programs—planned communication to specific audiences with many opportunities for feedback.

PROPAGANDA

Marketing and public relations are often accused of practicing propaganda. Before leaving this chapter, the concept of propaganda needs to be explored. Propaganda is usually thought of as lying by telling the truth—some facts are presented and others are obfuscated or ignored to present a very distorted view of reality. Certainly marketing and public relations messages, as well as many environmental messages are intended to present a point of view. Pratkanis and Aronson (1992) stated that "every day we are bombarded with one persuasive communication after another. These appeals persuade not through the give-and-take of argument and debate, but through the manipulation of symbols and of our most basic human emotions. For better or worse, ours is an age of propaganda." This is an important caution for environmental communicators, who often feel a moral commitment to the messages they craft. Propaganda used in pursuit of a social good is still propaganda.

However, there is a distinction between persuasive messages and propaganda. This distinction fails when dissenting voices are not allowed to speak. Environmental communicators must acknowledge other points of view. Environmental benefits will not be sustained if some publics are left out of the communication process.

SUMMARY

Marketing, especially social marketing, and public relations are powerful tools for environmental communicators. Both are planned approaches to communication to achieve specific objectives using certain techniques. Marketing and public relations approaches will be very effective in many situations. We encourage their creative application in a wide variety of natural resources settings.

REFERENCES AND FURTHER READING

Andreasen, A. R. (1995). *Marketing social change: changing behavior to promote health, social development, and the environment.* Jossey-Bass.

Bland, M., Theaker, A., & Wragg, D. (1996). *Effective media relations: how to get result.* London: Institute of Public Relations.

Boone, L. E., & Kurtz, D. C. (1999). *Contemporary marketing 1999.* The Dryden Press.

Caywood, C. L. (Ed.). (1997). *The handbook of strategic public relations & integrated communications.* McGraw-Hill.

Churchill, G. A., Jr. (1999). *Marketing research: Methodological foundations* (7th ed.). The Dryden Press.

Cutlip, S., M., & Center, A. H. (1978). *Effective public relations.* Prentice-Hall.

Dilenschneider, R. L. (1996). *Dartnell's public relations handbook.* Dartnell Corporation.

Fazio, J. R., & Gilbert, D. L. (1986). *Public relations and communications for natural resource managers* (2nd ed.). Kendall Hunt.

Goldberg, M. E., Fishbein, M., & Middlestadt, S. E. (1997). *Social marketing: Theoretical and practical perspectives.* Lawrence Erlbaum Associates.

Hendrix, J. A. (1998). *Public relations cases* (4th ed.). Wadsworth Publishing Co.

Kendall, R. (1996). *Public relations campaign strategies: Planning for implementation.* Harper Collins College Publishers.

Kotler, P., & Roberto, E. L. (1989). *Social marketing: Strategies for changing public behavior.* The Free Press.

Malhotra, N. K. (1999). *Marketing research: An applied orientation* (3rd ed.). Prentice-Hall.

Manoff, R. K. (1985). *Social marketing. New imperative for public health.* Praeger.

Pratkanis, A. R., & Aronson, E. (1992). *Age of propaganda: The everyday use and abuse of persuasion.* W. H. Freeman, Co.

Pratkanis, A. R., & Aronson, E. (1998). *Age of propaganda: The everyday use and abuse of persuasion.* W. H. Freeman, Co.

PRSA (Public Relations Society of America). (1999). http://www.prsa.org.

West, B., Sandman, P. M., & Greenberg, M. (1995). *The reporter's environmental handbook.* Rutgers University Press.

Young, D. (1996). *Building your company's good name: How to create & protect the reputation your organization wants & deserves.* American Management Association.

Zaltman, G., Kotler, P., & Kaufman, I. (Eds.). (1972). *Creating social change.* Holt, Rinehart and Winston.

Index

P

panel discussion, 110
paralanguage, 147
participant observation, 74
personal interviews, 73
personal space, 144
personality types, 113
persuasion, 6
physical movement, 144
placement, 186
planning, 45
players/stakeholders, 36
pluralism, 127
population segmentation, 55
positions, 37
Positive Mental Attitude (PMA), 141
"powerful others", 62
presupposition statements, 148
presumptions, 30
price, 186
primary literature, 26
primary motivation, 104
problem, 36
problem statement, 46
process, 77
product, 186
program evaluation, 77
projection, 136
promotion, 186
propaganda, 188
proxemics, 144
psycholinguistics, 147
public affairs reporting, 6
public relations, 185
public service announcements (PSAs), 91
public speaking, anxiety about 138
purpose, 84
purposes of evaluation, 71

R

reasoning, 32
receivers, 14
redefinition, 40
relaxation techniques, 141
reluctance, 79
resolving disputes, 173
revelation, 105
risk, 177
risk acceptance, 179
role of the media, 164
role playing, 109

S

sample make-up, 79
Satir modes, 114
"satisfaction-progression", 66
science literacy, 21
science reporting, reality of 168
science writing, 6, 91
scientific community, 7
scientific information, 21
scientists/engineers and the news, 167
secondary analysis, 77
secondary literature, 27
secondary motivation, 104
semiotics, 145
senders, 14
similes, 31
situational factors, 62
slides, 154
sociotyping, 130, 131
solutions, 37
source, 28
Speaker Apprehension Test, 140
spiral, 172,173
Sprouts, 56
stereotyping, 130
stretching, 141
structuring the presentation, 133
subtext, 143
subtyping, 130
summative evaluation, 49
support, 56
suppression, 127
surveys, 72
symbols, 145
symposium, 110